WE WERE YAHOO!

FROM INTERNET PIONEER
TO THE TRILLION DOLLAR LOSS OF
GOOGLE AND FACEBOOK

BY JEREMY RING

Post Hill
PRESS

A POST HILL PRESS BOOK
ISBN: 978-1-68261-578-2
ISBN (eBook): 978-1-68261-579-9

We Were Yahoo!
From Internet Pioneer to the Trillion Dollar Loss of Google and Facebook

Post Hill Press
New York • Nashville
posthillpress.com
Published in the United States of America

Dedication and Acknowledgements

Dedicated to Elijah, Levi, Eliana, Galit, and Stacey.

Elijah, your deep soul and thirst for knowledge always amazes me.
Levi, what you've accomplished and overcome at a young age is the inspiration that drives me every day.
Eliana, your heart and spirit widens my smile.
Galit, your humor and liveliness light up every room
Jessie Ring, no one ever had as many best friends as you. Your life was embedded into more hearts than you ever could have imagined.
Stacey, your unconditional encouragement during dark times was the fuel I needed to write *We Were Yahoo!*.

Special acknowledgements to: Jennifer Cohen, Billie Brownell, Anthony Ziccardi, Rob Buschel, Corey Friedman, and Drew Lanham.

TABLE OF CONTENTS

PROLOGUE

As the hours went by, I used the time to reflect on my adult life. I didn't intend to, but the thoughts that flooded my mind overwhelmed me. I initially thought about how my promising and successful life just a few years earlier turned into such a hellish soap opera. How did I fall so far that I was sitting in a cheap diner overdosing on vanilla milkshakes with sixty-year-old, barely dressed waitresses waiting for a madman with no teeth and horrible breath to appear for his four million dollars that I supposedly owed the mafia because of a bad stock tip? I wondered if any other young successful Internet millionaire faced similar plots. I was out of Yahoo! for only two years, but it felt like a lifetime.

At that moment, I was acutely aware of how depressed I had been for the past two years. I hated my life in Florida, I hated my job, and I just wanted life to be normal. But it couldn't be. It is not normal to be a part of growing one of the most iconic companies in the world. It's not normal to have made millions of dollars in my twenties. It's not normal to be thirty-two years old and have nowhere to go but down. But mostly, it's not normal to be extorted for millions of dollars by someone who introduced himself as the producer of the legendary Broadway show *Cats* and as the producer of Dustin Hoffman's and Al Pacino's next movie. This crazed lunatic mafia impersonator threatened to kill my family and me and stalked and tracked us to our five-star hotel hideout with mercenary bodyguards to warn me that he poisoned my sushi. Sitting alone in a cheap diner as I waited for him to pick up four million dollars was not normal life. It was severely f***ed-up!

Maybe it was all the milkshakes, but suddenly, my mood changed. I no longer felt terrified. I realized I hadn't had this much excitement in two years. I vowed to overcome my depression and decided to create a plan for the rest of my life.

Waiting for Jerry to appear was an ideal moment to determine my future. I drew a box with four squares. The top right square was to ensure I dedicated myself to spirituality. I'd had an epiphany. I had been given so much, but I never gave back. While employed at Yahoo!, I treated my years there as a blessing. After resigning and enduring depression and difficulties, I viewed Yahoo! as a curse. In truth, it was neither a curse nor a blessing: It was a test. I'd been provided all the resources, but did I have the strength, mind, and wherewithal to dedicate myself to creating a path to improve the lives of others? Up until that point I hadn't.

The second square was an extension of the first. I decided to form my own charity. I wasn't going to continue writing checks to large organizations where money is spent proportionally high on administration costs. I committed that any dollar spent went directly into the community of need. Whatever charitable organization I created had to have my vision where a need was most pronounced. I hadn't yet determined what the charity would support, but I decided I had to do something.

The third square represented dedication to my children. Both my sons were young, and while I loved them unconditionally, my career had held me back from being as active in their lives as I should have been. I was determined to give my children the proper life lessons and experiences, and ultimately raise them to be productive and compassionate adults.

The fourth and final square was a reminder of the Yahoo! legacy. It was the for-profit capitalism square. As much as earning money was a curse, it was also an addiction. I wanted to build a business, not just any business, but another world-changing business. Yahoo! had created a high bar, and

I wanted to exceed it. I wanted to create something that would change the world and allow me to get even wealthier while doing it.

I'd just created my four pillars of life: spirituality, charity, children, and capitalism.

I was reborn. I felt alive and exhilarated. Jerry Davis had done me a favor.

INTRODUCTION

January 10, 2000. Several executives just beneath the most senior management were having our regular lunch at Blue-Fin Sushi. We called ourselves the "Internet Crossfire Group." Our debates were legendary. No one let anyone finish their thoughts. The vilest curse words were integrated artfully into every sentence. Most arguments were about the state of internet companies—eBay, Excite, Lycos, Amazon, E-trade, E-toys, E-this, E-that, E-I-can't-f-ing-believe-how-much-f-ing-money-they-just-raised—out-of-control valuations of companies that had just gone public, and private companies with no business model, no prospect for revenue, but flush with millions of dollars to spend to build their brand so they too could "cash out."

This day was particularly animated. We had all awoken to world-shattering news. For 160 billion dollars, America Online had announced it had just acquired Time Warner. The most prestigious media company would now have the three letters "AOL" before its name. A ten-year-old company—whose only business was inside a computer—was taking over a company whose holdings included a vast magazine empire, the second largest cable network in America, multiple movie studios, professional sports franchises, and as valuable a collection of television and feature film content as anywhere on Earth.

Most of the talking heads—and the nation for that matter—were in disbelief and viewed this event as a seismic shift in all things rational. The internet had not just arrived; it had stomped and crushed traditional business overnight. To us, it was as if corporate America (as we had

11

known it for over a century) was overtaken by liberators, and the leaders of traditional business were walking through town with their hands above their heads on their way to a firing squad. On the other side of the barrel of the rifle were the young, inexperienced, arrogant, paper millionaires taking their turn to rule the new world.

For those of us at lunch, that day was a seminal event. We didn't view the AOL takeover as irrational or, for that matter, unexpected. It was possibly sooner than we had anticipated, but an inevitable outcome nevertheless. Our debate that day was one that only we in the entire world could be realistically having at that moment. Should we acquire Disney or Viacom, NBC or CBS, MasterCard or VISA?

Our company was five years old. All of us at that table were in our mid-twenties to early thirties. We were all multimillionaires. We were worth more than Ford, Chrysler, and GM combined. Hell, we were worth more that Disney, Viacom, and NewsCorp combined. Each of those great American brands could have been swallowed up by us.

We were Yahoo!!!

CHAPTER 1
Starting at the Bottom

The Ring family was raised in Canton, Massachusetts, a middle-class community. None of the residents were wealthy, and no one was particularly impoverished. My family included four children—three brothers and a sister, the youngest child in our family. Our parents had a wonderful loving marriage; I never even witnessed an argument. We grew up in a small town just south of Boston, the kind of town known for its strong New England work ethic. There was one main road, no McDonald's restaurant, three elementary schools, a middle school, and a 1988 high school graduating class of two hundred and fifty of the most homogenous student population imaginable. The town, Canton, was the historic home of Paul Revere. Henry David Thoreau briefly taught at the local high school where pre-revolutionary war homes and American flags line the main street. My mother left her job as an elementary school teacher to raise her children, all four of whom were six years apart in age. My father sold raincoats and umbrellas. He commuted on the Amtrak train Tuesday through Thursday each week to New York to sell his products in the Garment District. His office was in the Empire State Building, which always gave us a source of pride. His goods appeared in old-line department stores such as Woolworths, Sears, and K-Mart.

As kids, our fun was to hide high on a hill, above the main road, behind thick brush and bushes. As cars drove by at night, we would throw acorns at their windshields. The brave drivers would stop their cars, hear us roaring with laughter, and inevitably start chasing us. We would scatter, jump fences, climb roofs of commercial buildings, and hide in the woods, knowing we would never be captured or identified. This was our Saturday night amusement. As we got older, we would stand outside package (liquor) stores on Friday nights, begging for someone to buy us beer. The type of guy who always agreed to buy beer for us is the kind who hangs out at 3:00 a.m. in a Quickie Mart playing Keno. We always knew that as soon as we saw the special loser, we had our alcohol. We'd give him twenty bucks and tell him to buy us a case of beer, usually something very cheap like Geneseo or Shaefer. If we had a little extra, we would splurge for Coors or Budweiser. Once we had our case, we would head for the bleachers, where the rest of the high school students would have already set the bonfire and started their downward spiral from buzzed to drunk to passed out face-first. Usually around midnight, the police would shine their lights and get on their bullhorns. We would all run and jump a fence and be off. We knew the cops in our town were all fat and completely out of shape and there was no way at all that they could keep up with us. Unfortunately, the beer would all be left behind so the officers would gather up the full cans, put them in their car trunks, and then they would spend the rest of their weekends buzzed, drunk, and passed out face-first. Of course, many of these officers were the fathers of our friends, so we would wait until they fell asleep and just steal the beer back. We had a very circular and small world.

One night when I was working as a short-order cook, a group of friends stopped by at the end of my shift, just before the restaurant closed. In the next town over, there was an old, decrepit home. It was right of out of your worst nightmare. The trees in the front lawn that hadn't been tended in years; the branches hung so low that a good gust of wind would knock them off. Cobwebs hung from the leaves that seemingly hadn't been green in years. Hedges that were unpruned grew into the house. Its garage door looked as if a car had crashed through it. The outside had that dark goop that gets all over white aluminum siding hanging down the house. The gutters were hanging half off. This was the sort of home where you never rang the doorbell on Halloween because no decorations were needed to make it frightening.

The big rumor in town was that this house was owned and occupied by a cult of devil worshippers. There was no way we were going to allow devil worshipers to worship in our community. So a plan was hatched. We were going to drive to the house, get out of the car, and light bottle rockets and firecrackers right by their dilapidated windows. My job was easy. I was the driver. I was chosen as the driver probably because we used my car. It was a ten-year-old Chevy Nova station wagon. This car was a true classic. It had a huge bus-like stick shift in the middle. In the winter, I could never drive alone. Because it was rear wheel drive only, it could not go up an icy hill. Someone always had to be there to help push the car up the hill. The fan belt constantly broke, and the muffler was usually falling off and scraping against the road. We figured the best way to fix that was just to blast the stereo so we wouldn't have to hear the awful sound of muffler against the street. For someone like myself, having a car like this required learning at least the basics, or risk getting stuck in a massive snowstorm with howling winds and below-zero temperatures. The reason I was chosen to drive, however, was clear. All of us could fit in my Chevy Nova. The plan was simple: light the fireworks, drive off, laugh hysterically, end up at the package store, find someone to purchase a case of beer for us, meet at the field, and crawl home when the sun rose. It sounded like a perfect night for a sixteen-year-old.

Unfortunately, the plan was quickly derailed. It started smoothly. Everyone jumped into the Chevy Nova, four in front, four in back, and two in the way back. We successfully lit the fireworks. The laughter began. We hopped in the car, pulled out, and drove off to the package store to find the drunk. Unfortunately, just as we pulled onto Bay Road, the sirens blasted, and we were busted. The officers demanded we get out of the Chevy. I was terrified. Although I had definitely deserved many times to have been caught, more times than I can count, our actions led to a notice in the police log in the weekly newspaper. Never before had I actually been caught. I didn't know how to react. Fortunately, since it was a small town, they had no interest in charging us with any crime. They were, however, interested in scaring the "excretion" out of us, which they did quite effectively. The police hauled all of our asses to the station and threw us all in a clink for the night. They called our parents, with whom they were all friends. The cops and our parents hatched a plan that no one would be picked up until the morning so we all had to spend the night in jail. I can't say that everyone in the group was scared, but I was petrified. I made a promise that night that I've kept the rest of my life since. I will *never* again drive ten jerk-offs in a Chevy Nova station wagon to a house that is rumored to be occupied by devil worshipers.

It's clear that it's a long way from package stores, ten miscreants crammed in a station wagon, illegal fireworks, and jumping fences to "let's buy Disney," but that's where my life took me, and I clearly wasn't prepared. I didn't graduate from an Ivy League school. I had no business experience. I was hired at age twenty-five to be the first Yahoo! salesperson, and I hadn't as much sold a cup of lemonade from a lemonade stand. I had never sold before. We opened the entire East Coast operation out of my apartment in Hoboken, New Jersey. I was twenty-five, and I had already held at least six jobs, several from which I had been fired either for being lazy or stupid.

I graduated from Syracuse University in 1992 with a degree in advertising. I have no clue why I chose advertising as my major. I hated commercials and always channel surfed away from them as soon as they appeared. There were other majors I could have pursued. Syracuse was world renowned for its broadcast media department, with graduates such as Ted Koppel, Bob Costas, Sean McDonough, and dozens of other recognizable names. I could have studied journalism; today I constantly recognize classmates with bylines in the most prestigious print publications in the country. With the crossover of print journalists to television, I could have been a great sportswriter making millions of dollars a year giving my useless opinion on ESPN or a political journalist blabbering nonsensical comments on FOX News or MSNBC. (Of course, many have said I have a face for radio). Not me, however; for no apparent reason, I chose advertising. After graduation, I moved back home with my parents. I spent the next several months frantically searching for a job. Other than some temporary placements, nothing broke until November, when I received a call from Ogilvy & Mather Advertising in New York.

O&M was among the most prestigious advertising agencies in the world. Founded in the 1940s by David Ogilvy, the firm had amassed a client roster greater than any firm on Madison Avenue. Their clients included American Express, Ford, IBM, AT&T, Procter & Gamble, and dozens of other top global brands. They had hundreds of offices all over the world. I had blindly sent them my resume months earlier and was shocked and amazed when they called. The human resources manager who called asked if I could be in New York the next day for an interview.

Coincidentally, my father had a meeting in New York scheduled for the following day, so I was able to get to the interview. After interviewing with several executives, they offered me the job. It was my first official job as a college graduate, my first job in the "real world." I recall all the speeches during graduation from Syracuse when the speakers told the graduates, *It's cold and harsh in the world; stay home with your parents!* Not me, all I wanted was to leave home. I was tired of sleeping on a twin mattress with a creaky wood platform beneath it, with my late 1980s NFL cheerleader posters on my wall and bowling trophies from when I was twelve still on the shelves. This was my chance. I was going to live in New York. I was going to live on my own. I was going to get in trouble. Be as stupid as I had imagined stupid was. Learn to smoke cigarettes (which, thankfully, I hated), drink *real* beer like Molson, not Coors, meet girls, who would be impressed that I worked at the largest, most prestigious advertising agency in the world. Meet others in their early twenties and begin lifetime friendships. Go to wild parties that didn't begin until 1:00 a.m. I had it all planned out—until I didn't! There was one fairly significant problem: the starting salary! It was a measly nineteen thousand dollars a year. How was I going to live on nineteen thousand dollars a year in New York? Miller beer was out. Girls were out. Explaining that my starting salary was nineteen thousand dollars a year was not exactly a good pick-up line. I didn't even know how I was going to afford to take the subway to work each day. Knowing the challenges of surviving, I still knew that I had to forge ahead. The first order of business was finding somewhere to live. Analyzing my finances, I determined I could afford three hundred dollars per month in rent. I opened up *The Village Voice*, which in 1992 pre-internet days was *the* place to search for New York apartment listings. I quickly realized there wasn't much of a market at my price point. After a few days of researching, I was amazed that one was listed for three hundred dollars. I jumped all over it, called the landlord and told him I was on my way over. The apartment was in Park Slope, Brooklyn. While I had always imagined living in Manhattan, Park Slope was a beautiful, well-kept neighborhood so I wasn't too distraught that I'd be living in Brooklyn. When I arrived, I was thrilled to see a gorgeous apartment overlooking Prospect Park, one of the great treasures of New York. How lucky was I? As I toured the apartment and its three bedrooms, I took note that all the rooms had someone living there with no one seemingly looking as if they were moving out. I turned to the landlord and asked him what room I would be renting. Hoping he would point to one of the gorgeous rooms off of the hallway, I had to catch myself when he pointed to where I'd be staying. The living room had a small cul-de-sac, just large enough to fit a single twin mattress. My jaw dropped.

I asked him, "So, you are renting me the closet?" His response was what I should have anticipated. "For three hundred dollars," he said, "What did you expect?" I had been up, down, and through *The Village Voice*. I knew there was no other apartment I could afford. My parents were not going to continue to pay for my New York City hotel. I was really left with no choice. Pondering it deeply in my mind, I finally relented and agreed to go live in the closet.

The next week I officially started my new job at O&M. Even though my salary was minimum wage, I was still excited that I would be working in a professional environment for one of the most successful advertising firms of all time. I was truly looking forward to learning the business and was dreaming of a long and successful career in advertising. About a half-dozen of us started that first day. We all began as assistant media planners, responsible for providing support to the media planners whose role was to determine the times and programs the television and radio spots would run, in which magazines the print ads would appear, in what cities to purchase billboards, and anything else that had to do with the planning of an ad campaign. Although I dreamed of being in the creative department, I had neither creativity nor artistic ability, so my only fallback was the media team.

That morning we went through the typical first-day activities such as hearing about insurance and benefits, signing for the company handbook, learning where the bathrooms were, being assigned to a cubicle, and last, discovering which account we would be working on. Most of the new hires were assigned either to the IBM or American Express account, which just happened to be the two most prestigious accounts in the firm. When it came time for my assignment, I was told to report to the Pepperidge Farm team. I felt as if I were picked last on the playground. My new colleagues on IBM and American Express were the popular kids in class. The girls were gorgeous and the guys were preppy and funny. I was none of that. I was short, skinny, and had the personality of an empty beer bottle. My fears were realized that first day when our vendors splurged for Knicks tickets and dinner both for the IBM and American Express teams. At the end of the first day, I left the building alone. It was dark, cold, and snowing, and all I had to look forward to was returning to my closet.

As the weeks passed, it became more apparent that I had been screwed by being assigned to the Pepperidge Farm account. O&M was a company with hundreds of offices across the world with more than fourteen thousand employees. It was the most prestigious firm with the most prestigious clients and I felt as if I were the least relevant employee worldwide. My suspicions were not without some validation. I was at the lowest salary scale in the company. I had the dullest account in the company. I was about ten thousand layers down from the CEO, who would never have a clue who I was, and my direct boss was only making twenty-four thousand dollars a year.

I worked at O&M from November 1992 through March 1994. During that time, all the people with whom I'd started were promoted to media planner. I was stuck as an assistant media planner. The job was awful. No one but me knew how to use a computer. For those fourteen months, my only task was to take the handwritten charts handed to me and make them presentable using either PowerPoint or Excel. I was doing administrative work, and the problem I had was that the secretaries were all paid thirty thousand dollars a year. O&M had this great bargain with me. Make me the secretary but pay me like a Third World farmworker. I learned absolutely nothing about the media business. Never once did I meet with magazine publishers or network executives. I had no clue when and on what programs our spots were scheduled to air, and above all, I didn't care that "Pepperidge Farm Remembers."

It was obvious that I wasn't going to be promoted to media planner anytime soon. My time as an underpaid secretary prevented me from learning the skills I needed to be a full-fledged media planner. I would go home each night and pull hair out of my head because I hated my job so much. It wasn't challenging. No one gave a damn about me, and at a level where we were *all* entry level and no one had *any* experience, I was the least experienced assistant media planner in the entire industry. Something had to give. I either had to quit and find another job, or I had to be promoted, which even in my young, inexperienced mind, I knew would likely be impossible. I figured I'd try the Hail Mary pass and ask for a promotion. I made an appointment with the media director, who was the boss of the entire media department. He was an older gentleman who was aloof, but at least he knew who I was, which can't be said for the other 99 percent of the department. He was one of those executives who was afraid of a computer, so he often dropped his chicken scratch on my desk and asked me to turn it into something presentable. I walked into his office, not intimidated at all. I figured the worst that would happen was he would fire me. I knew there were other jobs such as being a trash collector, berry picker, doorman, museum security, or even rodeo clown that would pay at least twenty-one thousand dollars a year. I figured some of those jobs would likely be available. I promptly got right to the purpose of my visit. I demanded to know if and when I was going to get a promotion. I most assuredly startled him. You could see the wheels turning. He was struggling with how to answer me. I didn't need him to say much; his body language said it all. He was finally able to cobble together something barely coherent, but the message was clear. I was not ready for a promotion. I hadn't done enough to learn the intricacies of being a media planner. I thought to myself, "No s--t!" I had no real exposure to the business nor to a mentor who was guiding me. Another clear point was that he didn't want to fire me. Why would he? I was the cheapest secretary they had, and my understanding of Microsoft Excel and PowerPoint made me more than a commodity to him. Even though I shouldn't have been surprised, I was still livid. I even requested a transfer to the IBM or American Express teams. I was quickly rebuffed with the "there is no room" line. No room? Are you f-ing kidding me! Each of these two clients was spending hundreds of millions of dollars a year, and they didn't have room for a nineteen thousand dollar a year glorified secretary? No, this was clearly a case of the jocks not wanting to have to mingle with the unpopular kids. I thought that ended after middle school. I never imagined it existed in corporate America. I wasn't going to take this! I was going to do what any self-respecting just dumped upon total loser would do. I was going to go out to find a job that paid twenty-two thousand dollars a year.

Fortunately for me, it did not take very long. It's stunning how many companies will hire young kids who are eager to learn while being used and demand almost nothing in return. I even received an unanticipated bonus. The name of the firm was Publicis Bloom and this position paid twenty-three thousand dollars a year. Unlike Ogilvy & Mather, this was a very small agency. They had only two floors of a building in Midtown. They were a small satellite office of a much larger French advertising conglomerate. I was under the radar, away from the massive Madison Avenue monstrosities. They had a nice, but not overwhelming, client list. They represented the Bacardi brands, some traditional packaged goods accounts, and several business-to-business clients, who spent significantly in trade publications. Just before my arrival, they had brought on a new client, the CIT Group, and I was assigned to work on this new account. CIT was a Fortune 500 company that had been founded in the early 1900s to provide loans for small business and middle-market lending. The company had billions of dollars in revenue. However, it was actually a last resort loan company. The interest payments were inordinately high. They were my client, and they were a major corporation, but I viewed them as one step above a New York check cashing store and two steps above getting a loan from Tony Soprano. There's little to report on this very quick and minor stop through the underworld of the New York advertising mafia, other than to say I was fired after six months simply for sucking at my job. As I discovered, the only responsibility for this account was to buy print ads every day in newspapers nationwide. It wasn't as if we were buying full-page, four-color spreads in *USA Today*. We were buying quarter-page ads in local newspapers in cities like Peoria, Racine, Cheyenne, Mobile, Elmira, Allentown, East "do u really exist," West "r u f-ing kidding me," North "so you married your sister," and South "your wooden teeth are rotting." I couldn't handle it. I should have snapped but instead, I just retreated. I'd show up every morning at ten o'clock. At noon, I would disappear for a two-hour lunch. I'd reappear at two o'clock and play Minesweeper and Solitaire on the computer until five o'clock when I'd run out the door faster than Carl Lewis in the one-hundred-meter dash. I never took the time to make my cubicle even *look* as if someone worked there. There were no papers on the desk, no pictures on walls, just a big ass-print on the chair from hours of mindless Solitaire on the PC.

Now, to be clear, I wholeheartedly deserved this firing. *I* would have fired *me*; a six-year-old would have fired me; the guy who came by the office once a week to shine our shoes would have fired me; the old man in the food truck would have fired me. When the head of the media department called me into her office, I halfheartedly wanted to ask her, "What took you so long?" If it had been me, I would have just pointed toward the door, told me to get out, and been done with me. I did no work. I had no belongings. There would never even need to be a record that I had ever worked there. To my great luck, I wasn't the one firing me. I was fired by someone who had a heart. The head of media was a woman who had obviously had been in the business forever. She was from the old days of *Mad Men*. She looked as if she partook in the fun about a thousand times too many. I'm sure her memory of the seventies was a fog, but considering her history, she made me the kindest and fairest offer I'd ever received. I was fired in late September, and the company agreed to pay me through the end of the year. Because my salary was so pathetic, I hadn't saved as much as the cost of subway token. With this unexpected gift, I had time to launch a real job search and not be forced to take a job as a bike messenger or museum guard. They were paying me to leave. I always knew what severance was, but I'd never heard of it being offered to a twenty-three-thousand-dollar-a-year, low-level employee whose biggest claim to fame was owning the Minesweeper record on the company computer.

By the next day, I had launched my job search. Although I wasn't going to settle for some dead-end glorified secretary role, I was nevertheless anxious. I was hopeful that, if I found an opportunity quickly, I'd have two salaries for the remainder of the year. I did, in fact, go on several interviews, many of which either turned me down outright or it was clear that the position simply sucked. My first interview was with Young & Rubicam, or Y&R. Y&R was a clone to Ogilvy & Mather. It had hundreds of offices, thousands of employees, and the most prestigious clients. The men wore suits and carried briefcases. It was row after row of cubicles. Walking through the quiet hallways, you could hear a fly fart. My impression was that the rank-and-file staff needed a note from management for permission to speak. I had flashbacks of O&M. The hair on my back was standing, the hair on my arms was crawling, the hair on my head was regressing. My lunch was two-thirds up my throat. In my mind, I was calculating what would be worse, selling tokens in a New York City subway token booth or working as a media planner at Y&R. Fortunately for me, I wasn't given a chance to work at Y&R. After interviewing with several executives, I was told a few days later that I wasn't right for the position, which was a decision that would come back to haunt them later.

I had several more interviews. It seemed as if I were walking in a dark forest, where all the trees had come to life and the branches reached out in hopes of grabbing and crushing me. Nothing was working. I even tried to interview with a brokerage firm that offered me a position. Now, the fact that the job paid nothing, I'd have to make four hundred cold calls per day, and

I'd have to pay for the industry exams *myself* was just a tad more of a tipping point than the free coffee and copy of *The Wall Street Journal* they offered each day.

Just as I was losing all hope, my luck changed. A job placement recruiter who took pity on me called me to tell me about an advertising agency in Greenwich Village. The firm, she promised me, was young, exciting, and dynamic. She set up an interview for me for that afternoon. I went back to my apartment, put on a suit I had purchased at Men's Warehouse just so I could look my best during the interview process, caught the F train downtown, and was at the offices of MVBMS within the hour.

The second I got off the elevator felt like a magical moment. I stepped off and froze; I had to stare at my surroundings. The agency was two floors of a loft. In the middle was the coolest spiral staircase I've ever seen. Its stairs were glass, but as you stepped on each stair, a gray film appeared, turning the stair translucent. The main conference room had a towering mural of a three-ring circus. The walls of the office were all brick, giving a true downtown loft feel for the office space. There was a pinball machine as well as a foosball table. Most of all, everyone—men and woman—wore jeans and t-shirts. The whole environment was breathtaking to me, and after a short wait, I was led back to the interview. I interviewed with a stunning, six-foot-tall, thin woman. She was gorgeous in a very natural way, no make-up, straight long hair, not pretentious, and very happy about her job. We spoke for about an hour, and we instantly hit it off. The account I would be assigned to was MCI, which was a very hot long-distance company that had gained tremendous market share and was seen as a very credible threat to AT&T. The fact that, eventually, they would break every law in existence and much of their management team would end up in federal prison was of no consequence to me in 1994. At this point, the staff was considered, young, brash, and aggressive. Most important, MCI was one of the first real companies to embrace the internet—which forever changed our society and how we conduct business—as a platform. By the time I left the building, I knew not only had I hit it out of the park during the interview, but I *had to have* that job. When I received the offer of the position a few days later, a feeling of jubilation overcame me. For the first time in my professional career, a ball bounced my way. I was only twenty-four, but I had a strong sense that this was the break I needed. The job offered to me was media planner, with a salary of twenty-eight thousand dollars a year. Considering two years earlier I had been making nineteen thousand dollars a year, I was satisfied that at least I was moving forward financially. After moving out of the closet in Park Slope, for the past year I had been living with my brother on Ninth Avenue in the neighborhood called Hell's Kitchen. My bedroom was the living room and my bed was the couch. Although I had no illusions that I wasn't still poor, I had at least graduated from being "dirt poor" and virtually homeless. With my new raise, plus the severance money I was still being paid from my previous job, I was able to afford a real bedroom in an apartment in Hoboken, New Jersey. Hoboken was a great town for young people in their twenties who were just starting out in life and couldn't afford to live in Manhattan. The only downside was riding the urine-stenched Red "Crapple" (Apple) bus to and from the Port Authority bus station each day.

I couldn't contain my excitement on my first day at work. I had hidden my Men's Warehouse suit in the back of the closet (figuring I would never need to wear a suit again), and I put on a pair of Gap jeans and a Boston Red Sox t-shirt. I was ready for my first day. I was even assigned a real office, not a cubicle. The office was small and nondescript, but I'd never had a door at work before. Come to think of it, I didn't even have a door in any of the places I lived in New York.

Little did society realize just how transformational these days were. Just a little over a year earlier, a twenty-year-old entrepreneur named Marc Andreessen had left the University of Illinois to develop a company and a product called Netscape. Netscape was the first consumer-centric web browser. It was the first window into the world that would change the course of life for almost everyone on this planet. Initially, few established businesses took advantage of this vast opportunity. Online services such as AOL, Prodigy, and CompuServe were trusted platforms, but they were closed environments that provided basic news, finance, sports, and entertainment information. The fact that many AOL users visited their service for the purpose of participating in adult (porn) chats was never discovered by the old-line companies. An open environment such as the World Wide Web? Stay away from that. It's just porn, bomb-making kits, and message boards filled with George Carlin's "seven words you can never say on television." Or at least that's how I imagined firms such as O&M and Y&R were describing the internet to their clients. In their backward, stale, nineteenth-century minds, *of course* they would steer their clients away from a new medium. They drew storyboards using genuine artists. Technologically speaking, their creative teams were as sharp as a marble. No one within their organizations possessed the capabilities to develop websites.

MVBMS was different, and we had the perfect client to embrace this new dynamic world. For us, this was the game-changer, and we were going to be the leaders. Both MVBMS and MCI would establish me as a twenty-first-century thinker and technological leader. While traditional businesses weren't embracing the online world, a cottage industry was being developed by early leaders in the space. *Wired*, which launched around this time, was the first publication to feature a digital as well as print version of their magazine. The title was *Hotwired,* and very intelligently they offered businesses a space to advertise at the top of the home page. These early advertisements were conceived as "digital billboards." They were static, meaning they remained on the same page at all times and did not link to any specific website. Of course, few businesses yet had created websites to which the billboards *could* link. The first advertisers on *Hotwired* were MCI, Toyota, and Pepsi. Although it's been disputed who was actually first, there is no disputing that MCI was among the first businesses to embrace online advertising.

Each day, I'd arrive at work early, awake, and ready to tackle another opportunity. I'd never had this sense of euphoria at work before. My mind was racing. I was on the ground floor of an entirely new advertising medium, one that I clearly sensed would not only reshape the media business, but one that would forever alter how we communicate, find information, and purchase products. I pictured myself as one of a small group of individuals who was invited to a party introducing the most important new product in the history of the world. The power of the internet took over my very being. I even had a jolting sense that this window into a new world would shape the rest of my life. It was 1995 and a year earlier, my father had passed away. Had he returned to Earth after even a single year, he wouldn't have recognized the planet. Netscape launched, Yahoo! launched, and other search engines were springing up. Email was replacing voicemail as the entire internet began inching itself into mainstream society.

In the summer of 1995, the MVBMS team had a brilliant idea (though honestly, it wasn't a huge stretch). The *Hotwired* experiment demonstrated that ads could be delivered to individuals online. *Hotwired* was simply a print magazine that happened to be on the internet, and its ads were just static billboards. For the internet to demonstrate that it was a new medium, certain criteria would have to be met. To justify its value, we had to ensure that as a user clicked on the banner, it would transport them directly to the MCI website, and then implement an action, most likely a transaction. It was the cleanest and most efficient means to validate an advertiser's campaign. Its success would revolutionize the advertising and marketing industries in ways that at the time were unimaginable.

CHAPTER 2
What Was Yahoo!?

Implementation of the concept required a website that was willing to work in partnership. The largest website using the criteria both of traffic and prestige was a small start-up out of Stanford University. Jerry Yang and David Filo were two graduate students in the electrical engineering program. To generate some excitement in their lives, they did what any bored Stanford University Ph.D. students did; they collected all the existing websites—which at the time were counted in the thousands, unlike today, which are counted in the hundreds of millions—and in their infinite wisdom, determined they could categorize each one. Sounds enthralling! While my claim to fame was to stand outside of a package store searching for anyone who could buy beer for sixteen-year-olds, these guys were determining that websites that detailed multipole expansions in the study of electromagnetic and gravitational fields at distant points belong under the Science and Technology category in the Yahoo! directory. Before coming up with the name, the service was simply called "Jerry and David's Guide to the Worldwide Web."

The fact that in just a few short years their geek hobby would evolve into one of the most valuable information tools on Earth was not necessarily the result of a power of great foresight. It just coincided with the explosion of the internet due to the web browser, the penetration of personal computers into the consumer market, and a lifetime supply of bandwidth that the phone companies were building.

Yang and Filo commenced their categorization project in early 1994, within a year after Netscape's birth. Not long after the initiative started gaining traction, they contacted Tim Brady, a former Stanford classmate who was attending Harvard Business School. Brady would become the first employee, after the founders, at Yahoo!, ultimately becoming executive vice president and head of production for the company. Brady's task was to determine if somehow his friends' little project actually had the legs to be a legitimate, revenue-generating enterprise. The business plan he developed strongly indicated real business potential.

While identification of the revenue sources was being debated, there was no doubt that traffic to the service was growing beyond Yang and Filo's wildest imagination. The growth was so explosive that the servers at Stanford could no longer handle all the page views the site was generating. The two young founders needed a back-up. Their rescue was provided by none other than Mark Andreessen, co-founder of Netscape. At the time, Netscape was the most popular web browser. It had over 90 percent market share, which by default made its home page the most heavily visited domain on the internet. Andreessen provided Yahoo! with the support and distribution it would need to become the top-ranked website in the world. Yahoo! was prominently displayed as the default search engine on the top page of Netscape's home page. This meant that a user's first visual into the online world was Yahoo! Any website to which a user wanted to navigate required using the Yahoo! search and directory. Overnight, Netscape's placement allowed Yahoo! to dominate the online world. (Note: Within two years, Microsoft would launch Internet Explorer (IE), which was the start of the first wave of the "browser wars." IE would eventually dominate the market. Netscape was acquired by America Online for four billion dollars, which proved to be an awful transaction.

Netscape never regained its traction. It was re-launched as a portal, which failed, and it was eventually shut down following the big Time Warner deal.)

Yahoo! was quickly becoming a global phenomenon. The internet was exploding, almost re-engineering the world overnight and Yahoo! was right in the middle of this global paradigm shift. Traffic was exploding on the service. A company party was even held when the site reached one million users (today, the Yahoo! service has over one billion users worldwide).

While Yahoo! was generating global buzz, I was back in New York at MVBMS contemplating the future of media and sensing that a titanic shift was occurring. All I thought about was this new medium. I knew it was going to shake the world, and I wanted desperately to be there on the ground floor. I had developed a routine with my online habits. About three times a day, I visited ESPN.com; as a sports fanatic, I craved any updates. I'd spend time on Time Warner's service. Before Time Warner was acquired by AOL, they launched a website that offered all of their magazines for free under one large service, called Pathfinder. The magazines included *People, Sports Illustrated, Fortune, Entertainment Weekly,* and dozens of others. *The New York Times* had also launched its own website, which allowed me to feed my hunger for rich, detailed news. ZDNet, the online publication of Ziff Davis publications, provided me with a forum for all the technology news that was required knowledge for any geek like myself. News groups and message boards consumed my attention throughout the day as well. Of all the websites I regularly visited, none grabbed my attention as much as Yahoo!, which started as a directory for websites, quickly expanded in its scope. It offered newsfeeds from Reuters, stock quotes, sports scores, entertainment, and other news features. The content was easier to access than the existing dial-up online services, and it was refreshed throughout the day.

To demonstrate just how archaic the Yahoo! service was initially, there was no search capability. This meant that no web crawler existed that could either search the internet to retrieve results or use the program to add the website to the categorization engine. Other search engines existed; AltaVista, which was created by a team of engineers at Digital Equipment Corporation (DEC) and Lycos, which was developed at Carnegie Mellon University, were true search engines that developed programs that returned search results based on relevancy, keywords, traffic, or some other metric instructed by their algorithms.

Yahoo!'s system felt more like Santa's Village in the North Pole, where the elves are busy building toys for children all over the world. They're singing, laughing, and dancing, and imagining the happy children on the other end of the very special, wished-for presents. The scenery for the Yahoo! elves was a tad different, but the outcome was the same. It was the fourth dimension elf world. Yahoo! elves worked in a back office with a cave-like feeling. Each had a mattress beneath a cubicle, which was more appealing than many of their apartments. The lights were very dim, and eyes were fixed, almost glued, to the computer screen for hours at a time. When the Web Surfers (their official title) were let out of their cage, their break consisted of a can of Jolt (Monster or Red Bull in today's energy drink world) and a game of Foosball. Of course, it usually took them several minutes for their eyes to get used to the light of the outside world.

Because Jerry and David built their service as a directory, it required actual humans to study each website by sight and then place it in the proper category. The categories were arts, business and economy, computers and internet, education, entertainment, government, health, and news. Each of the main categories was broken into sub-categories. For example, the government category could have a sub-category of state government, the next level could be a website pertaining to a governor, the level beneath that was legislation specific to a governor, and so on. The categories could go dozens of sub-categories deep. The Web Surfers' job responsibility was to study the website and match it with the proper sub-category. If one didn't exist, then the task was to create a new one that was rational within the system. The entire platform was no different than how a physical library categorizes books.

The Yahoo! Surfers were regularly adding thousands of websites a day to the directory. So each day, in fact, several times a day, I would click on the "New" and "Cool" buttons. While eventually millions of websites were added by web crawling technology, in the early days identifying and viewing new websites was still manageable. The Yahoo! Web Surfers would add a "Cool" icon next to the URL indicating it was a recommended website. The new button indicated they had just reviewed and categorized the site for the first time. I spent hours investigating the "New" and "Cool" worlds. I guess you could say I was one of the first individuals to fall prey to losing all sense of social reality, an affliction that has now engulfed almost everyone under the age of fifty.

Now that I was a full-fledged internet addict, Yahoo! was my primary brand—kind of like my Marlboros. I was the Yahoo! man, although I admit, I wasn't faithful to the brand. Often, I would sneak a puff of other brands, such as ESPN, Pathfinder, *The New York Times*, and so on…. However, I did a good job of hiding my unfaithfulness and was a strong Yahoo! unofficial spokesperson. Recognizing that Yahoo! was the perfect website to test this new advertising model that relied on return on investment metrics, I contacted Yahoo! and one of the three people working there (and to this day, I don't know who) forwarded me to a company called Interactive Media Sales (IMS). This creatively named company had just—within the previous weeks—been given the mandate to recruit five advertisers for Yahoo!'s initial banner advertising program. The first meeting I had with IMS was with a former advertising executive named Andy Pakula. Before joining IMS, Andy led new business development for the Deutsch Agency, and prior to that he had worked with several Madison Avenue firms. As luck would have it, Andy and I had the same vision. We both believed that internet advertising was more aligned with direct marketing than with traditional brand advertising. This was clear to us because the medium had the potential to be the most accountable marketing vehicle ever. Unlike television, print, or radio, the user could be tracked, and the results were immediate. Therefore, the business model most appropriate was one driven by Return of Investment (ROI). It also turned out that Andy had been trying to persuade Yahoo! to accept banner advertising. At first, Jerry Yang was reluctant, and he initially responded to Andy that Yahoo! would always be free. With their finances dwindling to near zero, Jerry was left with no choice but to accept advertising to fund his growing enterprise.

Several lunchtime discussions between Andy and I yielded a pricing metric based on number of impressions (i.e., the number of times the banner ad was viewed) and the number of times the banner was actually clicked upon so the user was redirected to the advertiser's website. These models were abbreviated CPM (which stands for cost per one thousand views, in this case of the banner ad) and CPC (cost per click). Fifteen years later, CPC, along with CPT (cost per transaction), remained the primary pricing metrics for online advertisers. Once we agreed on the model, MCI became the first company to advertise on Yahoo! We purchased the ads for twenty thousand dollars per month. This was primarily based on a twenty-dollar CPM, which meant that the plan expected the ad to be viewed one million times (1.0MM/20,000 = $20).

MVBMS created several banner ads to test different creative options for driving users to the MCI website. Ironically, while I was an expert, I thought, on all things Yahoo!, for some reason it never occurred to me that Yahoo! had fewer than ten employees. The brand was going global, it was watercooler chat, and there was no reason whatsoever to assume that it had fewer employees than the local bagel restaurant on 34th Street. Each day at 4:00 p.m. I would get an email from Jerry@yahoo.com with the click-through results of the previous day. On his end, Jerry tracked all the users who actually clicked on the MCI banner as well as the number of times it was viewed. As I've described, these were the Wild West days. We were one of the only advertisers on Yahoo! and banner ads were far from ubiquitous; MCI had the most prominent space on the page, and without any rules our creative never mentioned that the banner was an advertisement. Recognizing the obvious that "sex sells," we added the word "naked" inside the copy of the banner advertising MCI.

While I'm sure the fact that we had prominent positioning played a role, just a hunch by having the foresight to include the word naked, our click-through percentage shot through the sky. On some days, it wasn't unusual for 25 percent of the users who viewed the banner ad to actually visit the MCI website. Today, advertisers are thrilled if their click-through percentage is *.025* percent. Let's be honest; no one clicks on banner ads any longer. Of course, now there are rules. Your copy has to be clean and everyone knows the banner is an advertisement.

I've always believed that we deserved a prize for proving that society is positively sick. We learned that just by using the word "naked," one-quarter of the people would take the next step. Imagine if our copy had included any of George Carlin's "seven words." The MCI servers would have crashed! However, had we chosen those words and the user ended up at the website of a phone company, I'm sure there would have been a mass exodus to drop the service. Just imagine some horny, obese, middle-aged man, who can barely locate his privates, having hit the jackpot of a lifetime: anonymous free porn. That individual would have been a bit upset if their fantasy girl turned out to be a long-distance salesman.

MCI was thrilled with the results as were my bosses at MVBMS. However, the technology was still new and had kinks that had to be worked out. It wasn't uncommon for the MCI banner ad to drop from the service for a day or more. Yahoo! was still developing its ad serving technology. Whenever there was an issue, I would call Jerry, and like an immature and inexperienced young punk, I would blast him for sucking so badly and demand that he make good on the screw-up. I was used to this type of behavior from buyers. My father had sold raincoats and umbrellas in the New York garment district, and as shipments arrived incomplete or late, the buyer would have his head. Unlike my father, I was the one hollering, and Jerry was the one receiving the hollering.

Ironically, I thought Jerry was just an entry-level ad-scheduling employee whom I could terrorize, just as I had been a few months prior at O&M and Publicis. Keeping in mind that I had no idea that Yahoo! had fewer than ten employees, my assumption wasn't that off-base. To my surprise and embarrassment, I was very, *very* wrong. I remember vividly waiting in a doctor's office when I picked up a copy of *People* magazine. I also recall being amazed that it was a recent edition (I'm used to doctor's office magazines being at least three years old). Right in the middle was a story about Yahoo! My stomach churned and its contents two-thirds up my throat. Jerry was one of the two co-founders of Yahoo! This person whom I'd been harassing and making his life miserable because I could, was right there in *People* magazine. It was obvious Yahoo! was going to have a public offering soon, making Jerry an instant billionaire, and I was demanding click-through results from him. It would be comparable to calling Walt Disney if the line for the "It's a Small World" ride was too long, or calling Steve Jobs to fix my email problems, or calling Henry Ford if my brakes were squeaking. I was yelling at *Jerry Yang* over a banner ad.

Fortunately, as embarrassed as I was, Jerry never let on. Of course, once I realized who he was, my attitude toward him changed. The word that comes to mind is *reverence*. I kissed his ass, overcompensating for being the a-hole I had been. I would have probably resorted to calling him Mr. Yang had he let me, even though he was only two years older than I.

The Yahoo! advertising test was an unqualified success. As a result, I gained a reputation in New York advertising circles as one of the young leaders of the new medium. I'd only been at MVBMS for ten months but I was acutely aware of my newfound marketability. For the first time in my brief professional career, I was happy in my current job. Unfortunately, I couldn't escape reality. I was only making twenty-eight thousand dollars annually. I was engaged to be married, and I was in no position to pay for an elaborate wedding. My wife came from a Long Island Jewish family. Being from Boston, I had heard the term "Long Island Jew," but I didn't know what it meant exactly. Before long I figured it out. What it means is that the wedding planning actually starts the day after a girl's Bat Mitzvah (which, in turn, commences right after birth). By the time her daughter reaches the age of sixteen, a Jewish mother-in-law from Long Island has already chosen the flowers, band, catering hall, and items for the dessert table. All of this is arranged without any regard for cost. That's something either the father-in-law or—if enough luck is involved—the groom himself can pay. No such luck on *this* end. I could barely afford a subway token. In fact, to purchase the engagement ring, I had to redeem all of my savings bonds from my own Bar Mitzvah.

So I couldn't afford the wedding. I surely couldn't afford an apartment worthy of having a woman live there. A splurge on a romantic dinner was eating at Burger Heaven. Basically, I needed money, and I knew the internet provided me with a path out of poverty.

Although there were several leads for good positions, the one that most appealed to me was Interactive Media Solutions, IMS. Beyond the creative branding of Interactive Media Solutions, they offered the strongest upside. I'd be expanding the Yahoo! advertising program to advertisers all over the country. I would be working closely with Jerry, who by this time had become a close friend, and, although my salary was marginal, with commissions and bonuses I could make more money than I ever imagined. I was only twenty-four.

While it wasn't a difficult decision to leave MVBMS, it was, nevertheless, bittersweet. They were the first company for which I'd worked that empowered its employees, and in that environment, I was able to thrive. Also, the fact that IMS was my fourth job in less than two years gave me pause to consider if I was a job whore. Little would I know, in just a few short months I would shatter that record.

The offices for IMS were located in an executive suite in the Graybar building connected to Grand Central Station in Manhattan. IMS had three principals, but Pakula had the most advertising experience. He was a traditional New York ad person: emotional, arrogant, had been married several times, and was very focused. Of us all, he knew the business. In my mind, there was no doubt he was the most valuable. He was the most "real," and he had genuine industry respect. Andy and I became fast friends.

IMS, while hired to sell advertising space on the Yahoo! service, had a much broader agenda. Their plan was to aggregate dozens of websites and sell advertising across a network. This wasn't necessarily a bad idea. In the subsequent years, companies such as Doubleclick and Advertising.com were all valued at hundreds of millions to billions of dollars using this strategy. IMS failed because, while they were the first to embrace the network concept, they were the last to develop the technology required to serve millions of ads across a network of websites while allowing creative to be changed on the fly and extremely detailed reporting beyond anything that the marketing industry had ever imagined.

The principals were securing the sales roles for dozens of websites. Early sites in the IMS network included Lycos, supermodels.com, AT&T, the NFL, and of course Yahoo! For me, I didn't care about any service other than Yahoo! I spent 100 percent of my time populating banner ads on Yahoo! from prestige clients such as IBM, UPS, Procter & Gamble, Toyota, and other Fortune 500 brands. Early on, there were only two basic advertising packages. An advertiser could either purchase banner ads in a specific category (arts, entertainment, technology, and so forth) or they could purchase a banner ad that appeared above a keyword search. The most forward thinking company was IBM. It's not hard to imagine that the most-searched word on Yahoo! was "sex," and by a huge margin. IBM purchased ad banners displayed above the search-word "sex." Their banner appeared more than any other advertiser on Yahoo!

Not only was "sex" the most-searched word on Yahoo!, but the top one hundred words were also all sex-related words. I always laughed when I'd read lists in newspapers or magazines of the most-searched words. In the '90s it was always reported as Brittany Spears, Tiger Woods, sports, music, and travel. None of those names or words could even crack the top one hundred of the most-searched. We just never let the secret out. The distressing part is that Yahoo! had hundreds of millions of users, which meant that we represented all society. Yes, we truly are a sick world!!

Yahoo! was the easiest service to sell; I felt personally connected to it and I had this sneaky feeling if I proved to Jerry that I could sell and not just buy, then he'd eventually *have* to hire me. I had a plan of how Yahoo! was going to hire me and nothing was going to compromise it. I was fortunate that I worked in New York. The other IMS principals worked in Los Angeles, and only Pakula was based in New York. Our office in an executive suite in the Graybar building was very stale. The only furniture and office equipment we had were an old, hand-me-down desk, a phone, and a computer. It was 1994 and the Graybar building had yet to be outfitted with a high-speed network; as a result, I was representing Yahoo! and selling banner ads for them when my access to the internet was a dial-up service through (ironically enough) AOL, Yahoo!'s biggest competitor. When I complained that I couldn't effectively do my job on a dial-up service that crashed regularly, the inevitable response was that most users were on dial-up, so we needed to be as well. While they were right, most internet users in 1994 were on a slow-speed network but their real reason was because they were cheap and didn't want to pay for a high-speed connection.

Within the month after I started, Pakula had a falling-out with the other principals and was gone. It was inevitable because he was the partner with the most advertising experience of the three, and thus he believed he deserved more of the company, but they refused to give up any of it. This was fine with me. Now no one was overseeing my day-to-day work, so I could focus all of my attention on Yahoo! If the other guys had an issue, they were three thousand miles away, so I really didn't care.

Not long after Pakula quit, Yahoo! hired Anil Singh as their vice president of advertising sales. Anil's primary role was to oversee IMS, the third-party ad sales team. It was evident that the eventual goal was to bring all sales in-house. I was committed to doing whatever was required to impress Anil. When he eventually built the in-house team, there was no way I was going to be left behind. Anil could be intimidating. He was of Indian descent but had been raised and educated in England. He was brilliant and always steps ahead of anyone with whom he was dealing. When asked a question, he always paused before he answered. His pause wasn't meant to put people off; it was that he never spoke without first thinking of what he was going to say. One could practically see his mind churning. However, most people were highly intimidated by his thought processes, which he used to his advantage. He would never tell you outright he was smart and you were dumb; he just had a way of making you feel as such. In most instances, however, that *was* the case. I have always been fascinated by our desire to prove how bright we are, and we've all been part of meetings where everyone talks above one another. But not Anil—he had this amazing quality of being a fantastic listener. In meetings, he never interrupted and never seemed impatient, even if the discussion went on for light years. He would wait his turn, which was always at the end. No matter how long the meeting was, nor what was discussed and seemingly agreed upon, Anil would utter just a few words, make his point, and the final decision was whatever he said. No matter who was in the room, we—and the other side—always knew who the final decision maker was.

Combined with Pakula quitting, from the moment Anil started with Yahoo! I just gave up on IMS. On Anil's first trip to New York, I escorted him to IBM, where we negotiated our first million-dollar ad deal. While Anil ultimately closed the deal, I set it up, worked it for weeks, and deservedly got credit. Immediately, my plan was working and I was gaining Anil's trust. Also, on that first trip, I impressed Anil at dinner by giving him a roadmap of the entire New York advertising industry and all the companies that would be potential clients. He came from a technology background, but not an advertising background, so I was an important resource for him.

I made sure to call him every day with the progress I had made. Fortunately for me, I was successful. Of course, this wasn't difficult. Our brand was growing so quickly, and the excitement of Yahoo! was so overwhelming, that basically I was an order taker. A turtle could have sold advertising on Yahoo! at this point. But that didn't matter to me. As long as Anil believed I was out there selling and I was hitting my goals, I would be best positioned to be hired once a sales team was brought in-house.

I clearly had no interest in stability. I had worked at O&M for a year, at Publicis for a few months, at MVBMS for less than a year, and I had only been at IMS for six months when a start-up company called Starwave, based out of Seattle, called.

Starwave was a small company founded by Paul Allen, co-founder of Microsoft. They were a web content and development shop. Their claim to fame is they created and managed ESPN.com and ABCNEWS.com websites, which were generating tremendous traffic. I was probably representing half their web pages, and I personally visited ESPN.com countless times each day.

The reputation I had launched at MVBMS in the internet advertising space business increased exponentially at IMS. As the only person representing Yahoo! in New York, I had instant credibility and visibility. Although I had only been with IMS for less than six months, it wasn't uncommon for my phone to ring often with job opportunities. I turned them all down. My goal was to work for Yahoo! and I wasn't about to compromise that goal. Or was I? When Starwave called, my interest was piqued. There were several reasons I was intrigued. First, I was an ESPN fanatic. Second, owner Paul Allen's net worth was worth north of twenty billion dollars at the time, so I had confidence the company was solvent and secure. Third, as much as I tried to ignore it, I was getting tired of IMS pressuring me to sell awful properties, such as supermodels.com, which had a shelf life as a business of about two minutes. Last, the person whom I'd respected the most, Pakula, had escaped but he left me behind—there was no one for me there anymore.

Rich Lefurgy, who was senior vice president of sales at Starwave, contacted me. Lefurgy, whom I didn't know very well, seemed like someone who was much more interested in building his own brand in the industry and using his current position as just a stepping stone for greater endeavors. He founded the Internet Advertising Bureau, which is the group that even today sets the standards for internet advertising. He was a keynote speaker at conferences all over the world regarding online advertising and was generally recognized as the *de facto* leader of this new medium. Though I had built a nice brand for myself, I was only twenty-five, and had never so much as sold a glass of lemonade at a lemonade stand prior to working with Yahoo! and IMS. Looking back, I was probably jealous of Lefurgy, simply because Yahoo! and I were doing all the pioneering, but it's understandable considering how young and naïve I was that he had the title of "big deal" in internet advertising.

The job that Lefurgy offered me was actually very intriguing. Starwave was a Seattle-based company, and they needed someone in New York to sell for them. They wanted me to open their New York office. The base pay of sixty thousand dollars a year was significantly higher than IMS, but at IMS I could have ultimately earned more through commissions. The decision was agonizing. Do I choose Paul Allen, co-founder of Microsoft, or leaders who forced me to work from an AOL dial-up network? While that was an easy choice, others weren't. Would I be giving up on my dream to work at Yahoo!? This weighed most heavily. Although Starwave was a solvent company, Paul Allen would likely never take his company public, meaning that the upside wasn't as significant. Yahoo! would obviously become a public company and I knew I could make a ton of money. The problem was, I wasn't working for Yahoo!, I was working for IMS.

I decided that in order to make the proper decision, I would have to visit Starwave in Seattle to check out the company. When I arrived, my first meeting was scheduled with Tom Phillips. Tom was the senior vice president of Starwave and had previously co-founded *Spy Magazine*, one of the greatest satirical publications of all time. I arrived at the office at lunch time. After waiting for about twenty minutes, the entire staff came through the door, sweating and laughing and loud. They had used their lunch break to have an ultimate Frisbee game. I knew then I wasn't in New York anymore. In New York, our lunch breaks consisted of avoiding taxis that would drive onto the sidewalks, having homeless people throw their unsanitized and disgusting hands in our faces asking for change, and waiting thirty minutes on salad bar lines at the local delis, just hoping that you weren't going to buy a salad in the middle of a salmonella outbreak.

I spent the afternoon meeting with the product team and the engineers. All day, I was amazed at how current they maintained the website. If you were a sports fanatic, like I was, then you'd reached heaven. Being in the room where all of ESPN's features and headlines were being sent out to the online world was analogous to a three-hundred-pound man taking a tour of the Hershey plant and falling into the chocolate vat. I was sold. I *had* to work for this company. Even though I'd be in New York, I spend as much time as possible in Seattle. I imagined myself playing ultimate Frisbee during lunch, soccer after work, and at night, basketball in Paul Allen's personal home gym. The rest of the day would be spent selling sports advertising. I knew there was no way that Miller, Budweiser, Coke, Pepsi, Ford, and GM would reject advertising on ESPN's website. With commissions, I could make a killing *and* sell a product about which I was passionate.

Before I left, I agreed to accept the position. I was to start in two weeks. I had to go back to New York and find office space. Though I wasn't worried about telling IMS, I was, however, very concerned about telling Jerry and Anil. So I didn't.

When I returned to New York, I frantically attempted to close all the Yahoo! deals I could. I had worked on several agreements for months, and they were close to being signed. I essentially wanted to hoard as many commissions as possible before I left. My last week at IMS, Jerry called me and informed me he was in town and wanted to have dinner. By now, Jerry and I had become close friends. We had a pleasant dinner. I don't recall the exact conversation, but most likely, it wasn't deep. We probably talked about sports, restaurants, the stock market, topics that were relatively benign. What I do remember clearly is what happened after we walked out to the street. Jerry informed me that Yahoo! was about to file for its initial public offering. Jerry was going to become an instant billionaire and all the employees, who by this time had numbered close to thirty, were all going to be assured millions. Jerry then gave me a stern warning not to do anything stupid. At that moment, I wasn't sure what he meant. He spoke of us all as a family, and how we'd be very successful if we all stuck close together.

As I was walking home, I was livid. Here, Jerry was telling me about the upcoming Yahoo! IPO, and how everyone was going to get very rich and when he had the opportunity, he didn't offer me a position. So in my mind, it meant everyone but *me* was going to get rich. I was still at IMS, at least for a few more days. I was seething but confident that I had made the right decision to accept the job at Starwave. Screw Yahoo! and screw Jerry. I felt as if I were doing all the work and they were making all the money. Even at twenty-five, I found such injustice with that.

Rich Lefurgy and I agreed that I would spend my first week at Starwave training in Seattle. IMS warned me it was the biggest mistake I'd ever make. I really didn't care. There was nothing they could say to me that was going to convince me that I should work with them. They even offered a substantial raise, but I figured that I'd make just as much money selling advertising for ESPN content. My first day at Starwave started out relatively uneventful. I spent the morning taking care of administrative needs such as choosing my insurance plan and filling out forms. Standard stuff one does on the first day. At about 11:00 a.m. my life changed. IMS had informed Jerry and Anil that I'd resigned and was at Starwave. My phone rang; it was Jerry on the other end. Before I had a chance to say hello, he immediately jumped into a barrage of attacks most often heard at a Philadelphia Flyers hockey game. *How could you f'ing do this to us? Are you an f'ing moron? What the f*** were you thinking? I just had f'ing dinner with you and told you not to do anything f'ing stupid?* It went on for a while. For an executive of a prominent company and a former Ph.D. student at Stanford and a media sensation, Jerry used the "F" word in that conversation as artfully and tactically as I've ever heard. After about twenty minutes of being called a f'ing moron, he slowed to breathe. Without being offended, I told Jerry that I'd chosen Starwave because Yahoo! hadn't hired me, and I was sick of working *with* Yahoo! but not *for* Yahoo! He admitted that he and Anil had made an agreement with IMS that they wouldn't poach any of their employees. However, since I'd quit, I was essentially a free agent. Jerry then instructed me to quit Starwave and fly to California to start with Yahoo! I reminded Jerry that I had only been at Starwave for three hours and it wasn't particularly good etiquette to quit so soon. He told me that Anil would call me later that night with an offer. I agreed to listen.

For the rest of the day, I was useless. I went through the motions of training, but it was going in one ear and out the other. I wasn't absorbing anything I was being taught. That night, back at my hotel room, Anil did call. He started off the conversation by strongly recommending that I come back to Yahoo! He reminded me that this is where my friends were. I knew the product, and I could claim responsibility for helping Yahoo! to grow.

Then he made his offer. Ninety-thousand dollars a year, plus 3 percent commission, plus fifty thousand stock options. The Yahoo! IPO was set for two months later, and with this amount of stock options, I was assured of becoming an instant millionaire. Keep in mind that I was only twenty-five and three years earlier, I had been in a dead-end world, lower than the secretaries on the company organizational chart and struggling to live in New York on a nineteen-thousand-dollar-a-year salary. I couldn't say no. I had dreamed about this chance for a long time and everything Anil said made sense. I was going to go work for Yahoo! I was going to be the first salesperson hired, other than Anil. I was going to open the entire East Coast operations of the company. Although I didn't know how big it would get, I knew I was going to be part of something that would change and reshape the world.

After I had finished with Anil, I called Sharon, my fiancée. We were scheduled to be married later that year. I explained to her that I was going to resign from Starwave and work for Yahoo! She told me that I was insane, that no one resigns from a job after one day. She was so upset that she even had her father call me to change my mind. He was a good businessman, and when I told him the offer, he admitted that I made the right decision. He even called Sharon, calmed her down, and she came around quickly.

The next morning, I walked into Starwave and told Lefurgy that I was quitting. He was amazed. No one had ever quit on him in one day. To be fair, no one quits after one day anywhere. He kept his composure, wished me luck, and asked me to help identify a replacement for me, which I did. I hadn't begun selling for ESPN yet, so there was no damage control needed. I left Seattle, flew down to Yahoo!, and started there that afternoon. Three jobs in three days in three different states had to be some sort of record. On Friday, I worked for a company in New York. On Monday, I worked for a company in Seattle. On Tuesday, I worked for a company in California.

CHAPTER 3
We ARE Yahoo!

As I walked into the Yahoo! offices, Anil and Jerry were there to greet me. We embraced and immediately got to work. No training was necessary. I just started to work the phones as if the last seventy-two hours hadn't happened.

I returned to New York and immediately opened the first East Coast Yahoo! office—from my apartment in Hoboken, New Jersey. I was feeling overwhelmed with amazement. In a few short months, the company would go public. I was the first Yahoo! employee outside of the headquarters. I would play a pivotal role in growing the company. Several times I asked myself how I got here. I had a middle-class New England upbringing, I struggled to hold a job, and I barely survived on the salary I'd been making. But I did make it, and I did so by the age of twenty-five. I was with Yahoo! Now all I needed was a desk!

January 1997

Bill Clinton had just won his second term as president of the United States. The nation was happy, peace prevailed, the stock market was flying at historic highs, and every person under the age of thirty was dreaming of starting an internet company and becoming a billionaire. Indeed, many were! It was a time of lasting hope. Venture capital, the engine for start-ups, was investing money more liberally than a bank bailout. The government was only semi-dysfunctional. Republicans and Democrats disliked and distrusted one another, but at least they could appear in the same room. Regular citizens, who had struggled their entire life financially, were starting to invest in internet stocks with amazing results. The stocks were society's first unlimited ATM for everyone to generate quick dollars where every citizen seemingly could participate.

In mid-1996, Yahoo! began to introduce web properties that served the needs of local communities. There was Yahoo! New York, Yahoo! San Francisco, Yahoo! Chicago, and so on. These sites delivered content specific to those regions such as restaurant reviews, yellow pages, local shopping, and other pertinent local information. To coincide with President Clinton's inauguration, Yahoo! Washington, DC was introduced. To promote the new website, we were invited to the inauguration to demonstrate the service, and because I was based in New York, I was asked to participate.

It was a bitterly cold day with a temperature of zero degrees. The inauguration committee had set up a tent to showcase new technologies. The growth of the internet was arguably the hallmark of Clinton's first term. While it's debatable how responsible he was in its explosion, the fact that it occurred during his time in office gave him considerable influence in the technology community. Thus, having a technology showcase was a no-brainer. It demonstrated the president was serious about leading a new innovative economy and that he was in step with the generation of dreamers and entrepreneurs. He could take credit for the millions of new jobs that were created due to this transforming communication and information platform.

The tent had over a dozen companies showcasing their newest innovations. IBM, Oracle, Samsung, and other major established Fortune 500 companies were invited along with Yahoo! It was surreal that our two-year-old business was awarded the same platform at the presidential inauguration as the greatest brands in the history of technology. The cold weather drew masses to the tent, where inside was a balmy twenty-five degrees, allowing inauguration revelers a brief respite from inevitable frostbite outside. As hundreds of thousands of people jammed the National Mall to witness President Clinton and Vice President Gore take their oaths of office, the lines to get into the warmer tents seemed as if they were miles long. As people streamed into the tent, shockingly they rushed past the world-changing demonstrations of the great innovators to congregate around our table. While IBM was demonstrating the world's great new supercomputer, we had a table with a computer showcasing links to local Washington, DC, restaurants and bars. IBM had spent hundreds of millions of dollars in development; our service cost barely anything to create. IBM's computer required collective contributions from the greatest engineers on the planet; triple PhDs from MIT, Stanford, and CalTech. Our service likely could have been developed by two high school students.

This was the event where I was struck with a clear realization. After two short years, our brand had developed an iconic status. No matter how cold I felt, I couldn't help but be thrilled and awed at the masses of citizens who wished to witness our new local service. Usually, cold weather would have chased me indoors, but on that day, I'd never felt warmer.

* * *

All of Yahoo!'s employees were in their mid- to late twenties. Traditionally at this point, young people are just graduating from junior level roles into mid-level positions. It's very rare to see a partner in a law firm under the age of thirty. Reaching the title of corporate vice president can take twenty years to achieve. Physicians in their twenties are still performing their residencies. For the internet generation, thirty was considered old age.

Had I remained in advertising, I would have likely been a media planner awaiting an opportunity to be promoted to associate media director. I was on a ladder to mid-level management. Whether I was assigned to work on the IBM, American Express, or McDonald's account, my client exposure would have been limited to mid-level marketing associates at those organizations. Senior executives meet with senior executives, mid-level executives meet with other mid-level executives, and junior executives meet their junior counterparts. That's just always been the corporate tradition.

It can take years in the ad world to make even a middle-class living. The big dollars are reserved for senior management. To counter the industry's low wages, many young agency executives turn toward a career in media sales. It's more common than not for a media sales rep to have gotten their start in an advertising agency. Although sales can prove lucrative, too often their exposure is limited to meetings with agency media planning groups. Their only hope of securing a meeting with the client CEO is if the publisher or network president is present in the meeting. At best, their client exposure is almost always limited to mid-management.

At Yahoo!, we never followed the rules of traditional corporate protocol. After six months at the company, Anil gave me the go-ahead to start hiring a sales staff. I reached out to sales reps who'd called on me while I was working at the advertising agencies. I tried to recruit individuals from *People, Time, Newsweek, Sports Illustrated, Entertainment Weekly, Business Week*, NBC, CBS, ABC, and Fox. All of them had heard of Yahoo!, but they were not sold on its potential.

Sales reps at large media companies had access to unlimited entertainment budgets, assistants, and an infrastructure that served all their needs. Many corporations also offered car services, dry cleaning, shoe shining, executive dining rooms, and corporate credit cards. Annual compensation was several hundred thousand dollars a year even for moderate performers. On the other hand, all we could offer was my office in my apartment, purchasing your own office supplies, no perks such as free meals and car services, and a much lower starting salary. It wasn't enough to offer tens of thousands of stock options and the potential to change the world. I was turned down by everyone. I begged. I channeled Steve Jobs when he famously convinced Pepsi executive John Scully to be the CEO of Apple by asking him if he wanted to sell sugar water or change the world. I posed the question as: Do you want to sell pages in a magazine or do you want to change the world? I discovered there were very few John Scully-like visionaries, and I certainly was no Steve Jobs. It was discouraging, but understandable. In 1996, few people really understood the power of stock options. If they had, there would have been a mad rush to work for us.

Unable to attract experienced sales people with established Rolodexes from major companies, I was forced to get creative. I called friends whose only background in sales was selling magazines subscriptions door-to-door in elementary school to raise money for a school field trip. I recruited a professional poker player, an unemployed college friend who was fresh off overcoming legal troubles, and a recent acquaintance from Ogilvy & Mather.

Our inexperience turned out to be our biggest asset. All we cared about was closing the sale. We were ethical but aggressive. Even though a few of us had ad agency backgrounds, we disregarded the traditional sales process and went directly to the client. At first, we'd meet with marketing managers at corporations but we realized that they didn't have the authority to approve our proposals. We'd wait for weeks, and even months, for someone in senior management to give us the go-ahead. This was entirely inefficient, and we were impatient. Yahoo! had just gone public, we were under the Wall Street microscope, and we had to reach our quarterly revenue goals.

We determined our best strategy was to go straight to the decision-maker. We *had* to meet directly with senior management. Without hesitation or caring about the traditional sales process, we began to contact the SVPs, EVPs, and CEOs of major Fortune 500 companies, such as Coca-Cola, Ford, Procter & Gamble, Barnes & Noble, Pepsi, and so many others. We were surprised to discover that it was as easy to secure a meeting with senior management as it was to get a meeting with junior or mid-level management.

Most senior executives were thrilled to meet with us. The Yahoo! brand could not be ignored. Major companies all over the world were struggling to figure out their internet strategies and considered us experts. It didn't matter that we were in our mid-twenties and many of these executives were in their late sixties with decades of experience. We had unlocked the code to making money online.

We could walk into just about any executive office in America. We positioned ourselves as a business solution when all we really cared about or had the ability to do was to sell them ad banners and sponsorships. But we weren't just vendors selling ad space; we were *business partners* selling ad space. That was the main difference from meeting with a mid-level marketing manager or the CEO. Ad sales reps from other media had been positioning themselves as partners for years, to no avail. Not us, we were immediately considered equals.

As our stock grew, our brand grew, and as our brand grew, our stock grew. Each fed off the other. Every company in America wanted to partner with us. Everyone believed that if they had a relationship with us, it would improve their stock prices, and often it did.

By 1999, we had started pitching companies to launch joint ventures. The agreement consisted of Yahoo! developing a co-branded online service and the partner marketing that service to their customers. The service was built on the My Yahoo! platform. My Yahoo! was our personalized home page. Users could customize their pages with sports scores, stock prices, news, travel, games, and other features. The co-brand usually consisted of the partner company having a small brand placement below Yahoo! and a module for them to offer services, products, and deals to their customers. Virtually 90 percent of the page was dedicated to Yahoo! and 10 percent was dedicated to the partner. Companies paid us millions of dollars for this service with the hope that we could spin off the service from the parent companies and take the joint venture public, raising billions of dollars. K-Mart was the biggest example of this sort of partnership. Unfortunately, the market imploded before any serious consideration of an IPO was possible.

It always amazed me that twenty-somethings could sit with CEOs and have conversations where we were not only listened to, but were considered the experts. We convinced CEOs of companies such as Ford, Proctor & Gamble, Barnes & Noble, and others that they were at risk of becoming dinosaurs and their future depended on us. Of course, over the years, this proved to be false. No major company needed Yahoo! But back then, they all believed they did.

As for the salespeople who rejected my early overtures, many of them eventually called and begged to be hired. I admit, I thoroughly enjoyed telling them to pound sand. I wanted risk takers, not salespeople who just wanted to jump on the bandwagon.

* * *

It was spring 2000. Our East Coast sales force, now over one hundred salespeople strong—and not working out of my apartment—was having a quarterly conference at Amelia Island, Florida. Part of my role as director of all sales programs globally was to attend each region's quarterly conference. Even though I was now living in Silicon Valley, I always looked forward to meeting the East Coast team. Not only had I launched the region, but I'd hired many of the salespeople and we became very close friends. It was a typically hot and humid Florida day. We had planned to work in the morning and golf in the afternoon. I always teased myself that, pound for pound, I was the worst golfer in America. My father was a tremendous golfer, and he started my brothers and me at a young age. In high school during the summer, we would play thirty-six holes a day. I've played thousands of holes of golf in my life. We were taught that practice makes perfect. In my case, that's bull. The more I play, the worse I get. I actually peaked at golf at age sixteen, and I've gotten progressively worse each year. My game has regressed so much that I'd bet against me in a match played against my eleven-year-old daughter.

That morning, we started the meetings around 8:00 a.m. We had no clue that the rest of the day would change our lives forever. One of the great parlor games at Yahoo! was watching our stock. Witnessing stock increases ten or more points a day was routine. For those of us living on the West Coast, we would awaken to CNBC to see our stock had risen by twenty points at the market opening. I always felt as if I were fooling the world. How could a company that had only a few thousand employees with revenue a fraction of that of other S&P 500 companies have a market capitalization higher than any media company in the world? It wasn't as if I wasn't proud of our work, it just never made sense why we were so overvalued.

I find conference meetings as painful as having my Achilles tendon severed. It's comprised of panels of people whose only real job is that of professional conference-goers. The type of people all large companies have that do absolutely nothing to help drive real revenue. Often their titles are corporate "evangelist." I've always resented these "evangelists." They have no accountability and talk as if they're experts. They're only experts at the free buffet offered at the lunches at which they speak. For some unknown reason, the morning session of the conference included several evangelists from our largest clients, E-trade, Target, Pepsi, and so on, giving a talk on the future of online advertising from the needs of the big brands. Normally, I'd be asleep before the moderator asks the first question, but on this morning my eyes were opened as wide as the Grand Canyon.

The stock opened that morning up ten points, which wasn't uncommon. Like all of us, I had my computer on my lap. I was logged onto Yahoo! Finance, checking the rest of my stocks. Every few minutes, I refreshed the stock price; all you had to do as a user was hit the refresh button on your keyboard. That morning the stock increased at a very steady pace; ten points at nine o'clock, fifteen points at nine-thirty, and by ten o'clock, twenty points ahead. While this wasn't unusual, it was always gratifying.

A large part of the compensation package at Yahoo! consisted of stock options. A stock option is not an outright share; rather, it's just an option to buy the share at a set strike price. The strike price is set the day you start, or the day you are awarded another round of options, often as an annual bonus. Yahoo! had split its stock multiple times in the past few years, meaning the strike price was cut in half at each stock split. The strike price for my options was about one dollar, mainly because I was hired before the company went public and before the stock price had risen through the sky. Someone who was hired in early 2000 might have a strike price of over two hundred dollars per share. Thus, the value of the shares was minimal, and if the stock price ever went below two hundred dollars it was worthless. Having a strike price as low as I had meant the stock would have to fall below one dollar for the stock to ever have zero value. With Yahoo!'s stock trading at over two hundred dollars that wasn't a possibility.

If the stock was having a particularly good day, our stock would usually settle up twenty to thirty points by market close. On a bad day, it could swing that much on the low end. On good days, the first Yahoo! employees would make tens of thousands of dollars or more on paper.

This day, something odd was occurring. The stock just kept rising and rising. By around eleven o'clock, all of the Yahoo! employees figured out what was going on. Our stock had already risen thirty-five points. The panelists were talking to themselves. These evangelists all believed in their brilliance. We should be honored because they gave us the privilege of being present for their words of wisdom. If there had been mirrors in the room, they would never would have left. None of us, however, cared. I was sitting on the other side of the room from my friend Chris, who was instant messaging me nonstop with the next update. It mattered little to him that I was looking at the same numbers. By lunchtime our stock had increased forty points. At lunch, we were all in disbelief. Something special was happening. We didn't speak; I think we were secretly calculating in our minds our net worth. There was a lunch speaker. I have no idea who was speaking, and I, along with everyone else in the room, didn't hear a single word that he said. We were all logged onto our computers, hitting the refresh button.

By two o'clock we were back in the conference room being subjected to additional speakers. If any of them had had a heart attack on stage, not one of us would have noticed. We were too busy watching our computer screens. Fifty points! Most stocks never reach a high of fifty dollars in their entire life-cycle; we were achieving this in five hours. Three o'clock—sixty points! I was almost comatose, other than wearing the biggest grin I've ever had. Finally at four o'clock, with the market closing, one final hit of the refresh button. There it was: seventy-five f'ing points!! Our stock had increased seventy-five dollars per share in a single day. The market value of our company was worth more than one hundred billion dollars, on revenues a tiny fraction of that. All I and others could do was put our hands over our faces and laugh. It's ironic that on a day that thousands of Yahoo! employees made more money than they could ever dream of, it was probably the least productive day in the history of the company. Jerry and David were billionaires several times over.

Our senior team was worth hundreds of millions of dollars, and everyone else had a value of tens of millions. The fact that it was only on paper and had little real life value and could have been Monopoly money was irrelevant. Five o'clock golf? Not a chance! How about the red-eye to Vegas, baby!

CHAPTER 4
Unanticipated Redemption

In early 1995 I was twenty-three years old and so poor that I struggled to save enough money to afford the subway at the end of the month. I had just been fired from Publicis Bloom and was in the midst of a job search when I interviewed at Young & Rubicam Advertising. Y&R was one of the oldest and most prestigious advertising agencies in the world. Established in 1923, their clients have included major brands such as Sears, Panasonic, and Mazda.

During that time, I interviewed with at least a dozen New York advertising agencies. Having had an awful experience at Ogilvy & Mather, it wasn't my dream to go work for another large agency with a similar corporate culture and similar hierarchy. The only conceivable differences were the salary, an increase to twenty-eight thousand dollars a year, and the position, media planner rather than assistant media planner. At Ogilvy, I was the absolute bottom of the company, below the mail room and cafeteria workers. The role at Y&R was a step higher, but still in the bottom three percent of the company.

While I was hesitant to work for another large firm, I chose to interview because I wasn't in a position to reject investigating every possible opportunity. A headhunter arranged for me to meet with the associate media director and group media director. As I was escorted to their offices, I was taken aback by how sterile the environment was. At least at Ogilvy, there was life in the hallways and employees gathered at makeshift meeting points. At Y&R, it was eerily quiet. An elephant could not have heard a sound. The people in their cubicles and in the offices reminded me of the living dead. I'd prefer solitary confinement. At least in prison, I could hold a more interesting conversation with myself.

The interview was uneventful. I answered the questions softly, afraid anything above a whisper would compromise the vow of silence in the rest of the building.

I've always subscribed that the most effective interviewing style is simply engaging a prospective employee in casual discussion. Establishing experience is fairly simple. Determining whether or not the person is a fit for your organization is much harder. I can capture critical factors, such as organizational skills, intelligence, and integrity through conversation, but in an office environment no qualification is more crucial than strong social skills.

The interview questions that day were standard BS. What are your strengths and weaknesses? Whenever I get those questions, I halfheartedly want to answer that my strength was that I wasn't either of them and my weakness was that I actually appeared for the interview. Lacking confidence, I gave the typical off-the-shelf responses. My strength was I work well with others and my weakness was I was too much of a perfectionist. Meanwhile, the entire time, I'm thinking if this is me in twenty years, please shoot me.

Ultimately no job offer materialized, which surprised me. I did my best impression of a half-dead future bureaucrat who clocked in at eight o'clock a.m., had no opinions, and would memorize the company fight song. It was apparent that a large, stale company wasn't for me.

As we age, it's natural and healthy to reflect on the great fortunes of our life. For me, being rejected by Y&R has to rank near the top. The events of the next twelve months were dizzying and life changing: being hired at Messner Advertising, getting early exposure to the commercialization of the internet, developing the internet advertising model, being hired by Starwave, and finally being hired at Yahoo! Had I gone to work for Y&R, I would have either been fired soon after starting or been on a management track to the living dead. At Messner I was empowered; I wore jeans and t-shirts, and I loved going to work every day. At Y&R, I would have been stifled, dressed in a suit and tie, and I would have dreaded going into the office every day.

Companies hire advertising agencies for their creative and strategic experience. Agencies develop ads, recommend the most appropriate times for the ads to appear, and recommend mediums that most effectively and efficiently reach the defined target audience. Lastly, agency media buying departments negotiate costs on behalf of the client. Essentially, the advertising agency acts a one-stop shop for all of a company's external marketing needs.

Part of the allure of hiring an advertising agency is that they are responsible for meeting with the hundreds of sales reps from television, print, radio, and outdoor companies. The client rarely had direct exposure to the vendor, until Yahoo!, and we changed that entire paradigm.

Major companies, such as Proctor & Gamble, Pepsi, and Coca-Cola, which spend billions of dollars on advertising annually, were negotiating marketing programs directly with Yahoo! Advertising agencies revenue is driven from acting as the agent for the client. Ad buys are marked up by as much as 15 percent. The client was saving money working directly with us while the agencies were losing millions.

In the late nineties, there were valid justifications for bypassing an ad agency. They were slow to embrace the power of online advertising; they were not technologically equipped or experienced; and most significantly, we positioned ourselves to be more strategic to a company's broader goals.

A great television spot has several purposes, but primarily it brands a company in the most positive sense. Gauging the return on investment can be complicated, though. A direct transaction often requires the customer to leave home to complete the purchase. The internet not only provides a customer the opportunity to purchase a product or service directly from home, but it allows the client to pay only when a transaction is completed. Traditional advertising agencies advertisers viewed themselves as artists, while at Yahoo!, we presented ourselves as an extension of the company's sales force.

With millions of dollars in lost billings, agencies were forced to re-examine their business models. Many launched internet-only divisions. They hired engineers and eventually became highly proficient in leveraging the sales power of the internet to its full potential. Targeting internet users for sales and marketing efforts has advanced considerably. A user visiting www.statefarm.com and then reading sports scores on ESPN is as likely to see a message selling home insurance as they are an ad for Nike. A user who visited the Johnson and Johnson website first and then checked their stocks on Yahoo! Finance is likely to have a pharmaceutical ad appear on their page. It's not just banner ads; advertising firms use granular targeting to optimize results on a search engine. The entire industry has become highly scientific, and the agencies have wisely installed themselves foremost in the process.

By 2000, valuations of tech companies were out of control. Start-up businesses with small revenue and no profits were being acquired for billions of dollars. Following American Online's acquisition of Time Warner, boardrooms of old-line businesses frantically debated whether to acquire, partner, or be acquired by a twenty-first-century digital company. The amazing power of the internet caught corporate America completely off guard. Without warning, Amazon overthrew Barnes & Noble as the world's largest bookseller. Google challenged Microsoft's dominance, and eBay was responsible for the demise of hundred-year-old auction houses. Newspapers across the country were so spooked they offered their content for free, a decision from which they've never recovered.

Similarly, advertising agencies were struggling to save their industry. While larger agencies were buying small start-up internet-exclusive ad agencies for tremendously high valuations, one business had grander ideas.

Most of Yahoo!'s experienced reps had come from traditional media companies. We often hired salespeople with magazine, newspaper, television, and radio backgrounds. All had close relationships with agency executives. They regularly heard from their friends the anger and distrust aimed at the large internet publishing companies such as Yahoo!, which, in their minds, were challenging their entire existence. Anil, sensitive to the fact that Yahoo! was being labeled as a difficult company to work with, instructed our sales teams to make a strong effort to reach out to the agencies.

As the head of the Sales Programs group, I traveled to New York to accompany the sales teams to visit customers. I began to meet regularly with advertising agencies in hopes of convincing them they could trust us as partners. Together, we could develop programs that met a client's goals. One of my first agency meetings was with Y&R. It had been at least five years since I interviewed with them. I'd given it little thought in the ensuing years. My life was a whirlwind of excitement. Losing a twenty-eight-thousand-dollar-a-year job was a distant memory that just wasn't that important.

Our meeting was with the agency president and the head of their business development group. If there was air that needed clearing, we wanted to start at the top. Ironically, if I'd been hired as a media planner years ago, I would never have been included in a meeting with the highest-level executives. They likely would never have even known I existed.

I was surprised to learn that the agency's chief financial officer was also at the meeting. The CFO is responsible for a company's finances, not its internet advertising programs. I quickly realized why his presence was required.

It had been widely reported in the newspapers and trade publications that WPP, the largest advertising and marketing conglomerate in the world, was close to acquiring Young & Rubicam, one of the last major advertising agencies to remain independent. Three large holding companies control most of the major advertising agencies in the world today. The companies include WPP, whose holdings included Ogilvy & Mather and J. Walter Thompson; Omnicom, parent company to McCann Erickson; and Publicis, owner of Saatchi & Saatchi. As fans of the AMC series *Mad Men* know, these are the most prestigious names in advertising.

After the introductions were made, I started the meeting by declaring our willingness to be strong partners. I wanted them to know definitively that Yahoo! was not the enemy. It was our intention to develop a relationship beneficial to both sides. I came prepared with several ideas to discuss.

Before I could present our concepts, their head of business development interrupted to share the Y&R vision for a future partnership. Uncomfortable with being swallowed up by WPP, she proposed that Yahoo! buy Y&R.

Y&R employed over sixteen thousand people, had dozens of offices worldwide, and billed close to a billion dollars. Yahoo! was barely a six-year-old company with fewer than two thousand employees. No one at Yahoo! had the first clue about running an advertising agency; in fact, my experience as an assistant media planner probably qualified me as among the most knowledgeable. I was completely shocked. I would have left happy if we sold some ad banner space.

I politely responded that I wasn't in a decision-making role with regard to acquiring billion-dollar businesses, but I was more than happy to set up a meeting for them to meet Anil and Jeff Mallett, the president of Yahoo!

Following AOL-Time Warner, it wasn't uncommon for major companies to engage in acquisition discussions with Yahoo! Our stock was valued at over one hundred billion dollars, and we were the highest of the high-flying twenty-first-century digital companies.

The difference here was this proposal demonstrated just how insane my world had evolved in no time. A few years earlier, Y&R had refused to hire me for one of the most junior positions in the company. Now, that same business wanted me to consider acquiring them.

I couldn't believe how far and how fast business had changed. Right after Y&R rejected me, I was part of the first team to embrace and implement internet advertising. The next few years witnessed explosive growth, unlike the world had ever seen. No more was this demonstrated than major, century-old advertising companies wanting to be owned by groups like us. Simply amazing!

CHAPTER 5
A Life We Never Imagined

By 1999, Yahoo! was established as one of the most important companies in the world. Jerry had appeared on the covers of *Newsweek* and *Business Week*. Jerry and David annually appeared on *Vanity Fair*'s top global media entrepreneurs list. Our quarterly stock announcements were hailed as a major media event. The company was growing by leaps and bounds. Keeping up with staff increases became challenging. Every day, there was another reminder, either through the media, client meetings, or water-cooler conversations, that we actually *were* transforming the world. We changed the way people searched for and discovered information. We altered the way people communicate, and we shifted the entire media landscape in less than five short years. Facebook, Twitter, and YouTube were years away from being invented. Google was a small nascent company, and Apple was teetering on irrelevance. In our minds, we were unstoppable and invincible.

Jerry and David, according to *Forbes* magazine, were two of the wealthiest individuals in the world and neither was yet thirty-three years old. One night it particularly hit me how far my new world was from my upbringing. Zod Nazem, the chief technology officer of Yahoo!, was hosting a poker game at his estate in Silicon Valley. I was invited along with Jerry, David, and a few other Yahoo! and non-Yahoo! executives. A wonderful personal chef served dinner. After dinner, as the game began, I realized there was a combined net worth of over twenty billion dollars at the table. The collective group could have acquired not just the New York Yankees, but probably all of major league baseball. I kept thinking about the number of Wall Street financial managers who would give their left leg and right arm to be present at that poker game. Interestingly, the group's cheapness was legendary. The game stakes were a quarter high, and the largest pot never exceeded twenty dollars. Ironically, the playing cards were kept in a Gucci case that probably cost more than all the evenings pots combined.

Not long after, we hosted our annual sales meeting in Monterey, California. Monterey is an exquisite, centraly located California seaside town overlooking the Pacific Ocean. It's just a few miles from Pebble Beach and home to some of the most beautiful golf courses in the world. Fifteen-hundred-square-foot cottages on the Monterey Peninsula start at over one million dollars. The area is comprised of Monterey, Pacific Grove, Pebble Beach, and Carmel-by-the-Sea. The conference was held at The Monterey Bay Inn, which is located right on Cannery Row, the area made famous by John Steinbeck in his 1945 novel of the same name.

Yahoo! sales executives from across the world arrived for the event. Reps from our offices in every major European and Asian country were present along with our Australian and Canadian colleagues. All of the U.S. sales representatives flew to Monterey as well. Several salespeople arrived a day early to play golf at Pebble Beach or shop in the stores on Ocean Avenue in Carmel. In total, there were more than five hundred worldwide "Yahoos" at the conference.

To say the mood was festive is an understatement. Our stock had split six times and was still trading close to two hundred dollars per share. Everyone in the room had a net worth of at least a million dollars. Sales assistants from our West Coast office drove to the meeting in their Porsches and Jaguars. I was responsible for planning many of the sessions for the conference. Anil sincerely hoped that work would be accomplished. One lesson learned is if you want to ensure a successful sales conference, don't give all the employees keys to the mini bar.

One night, word got out that there would be an after-party in my room. Unfortunately, my room was a standard size and could comfortably accommodate no more than five individuals before it was overcrowded. Within twenty minutes there were close to thirty people in the room. God knows what the net worth was, as both Jerry and David were there; I'd guess a combined net worth twenty billion dollars was present. All I could think about was that if there were a fire, no one would survive. There was enough wealth to fund the gross domestic product of a small nation that would have gone up in flames. Fortunately, another guest on the same floor had the good sense to call security. Even though it was my room, I was grateful to the hotel staff breaking up the mad scene.

When literally hundreds of young millionaires, with mini-bar keys, and away from home all coalesce at the same spot with no supervision, the results are not good. The hotel damages from the rowdiness added thousands of dollars to the final bill, and a not-so-subtle request was made for us never again hold our conference at The Monterey Bay Inn.

CHAPTER 6
A Storm is Coming

December 2000

I was closing in on my fifth anniversary with Yahoo! and the sixth year since we'd developed the initial advertising program. Unfortunately, the good times ended, and there was nothing subtle about the crash. The excitement we'd built over five years diminished in what seemed like five minutes. E-toys and toys.com burned. Then Globe.com imploded. Pets.com and that highly disturbing sock puppet just shriveled away. Global Crossing filed for bankruptcy protection. Already, the AOL-Time Warner merger was considered the largest corporate bust in history. Yahoo!'s stock, which had been trading as high as 250 dollars, was now trading below fifty dollars, and it wasn't through with its free-fall. The Dow, which had peaked earlier that year above fourteen thousand, was about to lose close to 50 percent of its value. The Nasdaq, where all the tech stocks were listed, decreased by over 80 percent. Venture capital dried up. Young, stupid internet millionaires who had bought homes in France, nightclubs in New York, and financed movies in Hollywood, all on margin loans, lost all their holdings when the loans were called. Everyday Americans who participated in the internet bubble lost their entire life savings. Retirement vehicles such as 401Ks and others were wiped out. Entrepreneurship, the heart of our nation, was effectively through. With capital dried up, innovation came to a screeching halt.

The bubble burst and with it, dreams were shattered. For a nation of individuals, who had never in history been able to generate money so quickly and with such ease, the collapse was catastrophic. It's harder to have had and lost than never to have had at all. A cliché, yes, but oh, so true!

Anil, always the smartest person in the room, gave a sobering speech during our summer sales conference, where we hosted more than five hundred sales executives from across the globe, whose theme was, "A storm is coming." I wanted so badly to disagree that the end was near, but for those of us so near to the ground, it was plainly visible; the good times had a short life. We knew clearly what the rest of the world would witness painfully within the year; the bubble had burst. Our awareness wasn't predicated on brilliantly predicting the future; rather, we had all the important indicators slamming us right in our guts, primarily all the web-based start-ups, spending millions of dollars with Yahoo!, all stopped paying us at the same time.

A third-grader in our position would have known the bust was approaching. Unfortunately, no one was in our position. Individuals nationwide continued to spend money that was essentially imaginary. So much of society was caught up in how high the stock market was rising, meaning in turn, how much their net worth was increasing, that we ended up as a nation with blinders on. Several of my colleagues and I were highly pessimistic privately, even if our public façade was quite the opposite.

I wasn't about to go down with the ship, so I sold all of my stock. I vividly recall the day I did. I should have been thrilled. It should have been among the happiest days of my life. In one day, I generated enough wealth to ensure the financial security for the rest of my life and my children's lives. However, it wasn't a happy day.

Selfishly, I knew the good times were over. The dream was over. Our rock star life was over. Insane, twelve-hour jaunts to Vegas were over. Having leverage in negotiations was over. Meetings with Fortune 500 CEOs all over the world were over. What entirely defined *me*—up to that moment—was over.

The stock would never recover, and I knew it. A new management team would soon be in place, and I knew it. The power of the iconic brand would soon begin to wear off, and I knew it. Millions of Americans would soon lose their life savings, and I knew it. Unemployment from the internet bubble crash would rise, and I knew it. Innovation would come to a halt, and I knew it. This generation's "Era of Good Hope" was over, and I knew it.

Knowing the storm was coming, there were times I wished I could have broadcast to the world to bail. *Get out now!* They'll be few survivors. In the end, it really didn't matter; no one would have listened to me. When individuals are making money quickly, the end is never in sight. No matter how loud I warned, no one would have wanted to believe me. I could have climbed thirty-thousand feet to the top of Mount Everest and hollered to the world; no one would have heard.

Having no illusions of my absolute lack of power and influence, I decided to turn to self-preservation. I had sold my stock; the company had, at most, two quarters left before Armageddon would ferociously attack. I needed an exit strategy, fast. Anil knew it, so he resigned, leaving me to take the brunt of attacks from executive management due to the lack of sales. What sucked was that there was nothing I could do about it. The companies that had essentially supported us with ad revenue were all filing for bankruptcy. It's like what would happen to the U.S. if China suddenly went out of business. We'd be screwed with absolutely no possible path for recovery.

I had been warning for years that we needed to focus less attention on closing big deals with start-up internet companies and focus more on smaller deals that would grow into larger programs over time with major companies such as Target, Wal-Mart, Pepsi, and other traditional advertising spenders. Unfortunately, being the most public of public companies didn't allow us that luxury. The demands of Wall Street mandated that, through hell or high water, we hit and exceed our quarterly numbers. The pressure was staggering. If Yahoo! had ever missed its quarterly numbers, there was a very real chance that we could cause an entire market crash, ensuring that millions of people around the world would lose their life savings. We lived with this pressure daily, and it's what nearly occurred when the numbers were missed—for the first time in *twenty-one quarters*. The combination of dot-com businesses liquidating their assets and Anil's resignation forced the realization that it was time for me to move on as well.

CHAPTER 7
Life After Yahoo!:
The Tom Bennett Story

Two years after I left Yahoo!, I received a frantic call from Chris Corrigan. Post Yahoo!, I was hired as the CEO of a small digital rights management company called Convizion. Corrigan was one of our employees, who had led our digitization efforts. He called me to tell me about an individual who appeared at our office earlier in the day. We met with prospective clients all the time, just at their offices. Our offices were small, less than four thousand square feet. So it was rare that we ever had a customer visit us. Although the company was based in Florida, our customer bases were Los Angeles and New York, so we spent much of our time on the road.

Thus, it was odd that a potential customer was there in the first place. While Corrigan couldn't recall the name of the individual, he was clear in his description. He explained how the man was tall and very thin, almost emaciated. He wore an old plaid shirt. He had a loud voice that carried deep into the back room, where our production team was housed. The man was bald, had no teeth, and had an awful odor that permeated the entire office. The man introduced himself to Corrigan as a movie and Broadway stage producer. He was face-to-face with Chris, who almost collapsed from the stench of the man's breath.

The man had researched our company and told Corrigan that the purpose of his visit was to purchase hundreds of thousands of dollars of film footage from the Convizion library for his next feature film. He presented a script and signed agreements with Dustin Hoffman and Al Pacino to star in the movie. One of the reasons I liked Chris so much was that common sense wasn't foreign to him. The smell of something odd was a deep red on his radar. He politely asked the man to leave, saying that he didn't think he could assist him. Although Convizion had provided archived footage to feature films in the past, Corrigan informed the man that we were out of the business of working on films, which was clearly a lie. The man didn't argue, and turned to leave. On his way out he ominously stated to Chris that he'd be back to speak to Jeremy Ring, and with that he left.

Corrigan and I laughed when he was describing the events to me. The thought of a horrible-smelling drunk with no teeth who represented himself as a major Hollywood and Broadway producer with signed agreements from Al Pacino and Dustin Hoffman as the stars was so unbelievable, it was hysterical. Chris was poking at me for being associated with such a loser. The thought of calling the authorities and reporting a wacko never crossed our minds.

Several weeks passed and the degenerative drunk was pretty much forgotten, when my life came crashing deep into a place I never could have imagined. The phone at my house rang and the voice on the other end asked to speak to me. I found this highly unusual. I've never given out my home phone number to anyone. It was most certainly unlisted, and I used my cell phone exclusively. The man introduced himself as Jerry Davis and he emphatically stated he must meet me immediately in a shopping center near my home. His sound was very distinct. He spoke with a high tone, but very convincingly and very demandingly. He also spoke very quickly as if he were on the run and had to deliver a message to me quickly. He had my full attention when he announced that Sharon, my two very young children, and I would all be killed if I didn't come to meet him right away.

I was petrified, but I decided to meet him. After all, he wanted to meet in a public place, so although I was anxious of the news he was going to deliver to me, I rationalized that I couldn't be harmed in public. If he'd wanted to kill me, he could have just come to my house. He obviously knew where I lived. I got in my car and drove about three miles to a Publix shopping center where he wanted to meet. I got out of my car and noticed an old, beat-up, rust-laden, blue Lincoln Town Car. The car looked like it should have been in a junkyard. We had agreed to meet in front of Publix, and as I got out of my car, a man appeared out of the beat-up car and walked up to me. He was tall, very thin, smoking a cigarette, bald, wearing an old plaid shirt, he had no teeth, and I could smell him a mile away. He pointed me toward the diner next door. I followed him to a table in the back of the restaurant. He sat down and instructed me to stay calm, that no one would harm me in a public restaurant.

He introduced himself as Jerry Davis. He handed me a piece of paper with his name written in pen and his cellphone number. At this point, my fear was inescapable.

Jerry started by explaining that he was a big deal in the entertainment industry. He caught me completely off guard when he declared that he was the main producer of the Broadway show *Cats*. He then bragged that he had conquered Broadway and was now setting his sights on Hollywood feature films. Jerry had just signed an agreement with Dustin Hoffman and Al Pacino to star in his new movie. Two thoughts quickly popped into my head. First, this was the man who had scared the hell out of Corrigan a few weeks earlier and second, how in the world is this man a major Broadway and Hollywood producer if he drives a rusted, beat-up, urine-stained car?

Jerry quickly moved beyond his biography. After a moment, he asked me to recall a meeting I had two years earlier with a gentleman by the name of Mr. Gennaro. I struggled to remember the name. Finally, I admitted I couldn't and asked where and under what circumstance we met. According to Jerry we'd crossed paths two years earlier at the Marriot Marquis Hotel in Times Square in New York. Supposedly, after a dinner, we met in the lounge and began a long discussion regarding Yahoo! Jerry explained that the discussion turned toward our high-flying stock. At that time, the stock had yet to tank. He saw that I clearly had no recollection of the conversation so he graciously filled in the rest. He continued, saying that during our discussion I strongly recommended our stock. Based upon that tip, Mr. Gennaro bought millions of dollars' worth of Yahoo! shares. It wasn't long after that the stock began its quick slide down a cliff. In truth, it fell from a high of 250 dollars to a low of eight dollars. I was listening intently. My eyes were locked on Jerry as he was talking.

He paused then he dropped a bomb. Mr. Gennaro, unbeknownst to me, was the head of a major New York Mafia family. My stock tip had cost him four million dollars, and unless I paid back every lost dollar by tomorrow, I would be killed. I turned yellow. A look of disbelief and fear crashed over me. I was terrified. Two years earlier, I was on top of the world and now I was at a table with a toothless, smoking, skunk-smelling, Broadway producer, friend of Al Pacino and Dustin Hoffman, drunk, old man telling me I was about to be killed by a major Mafia Don. I couldn't in my wildest dreams have imagined this scenario.

My reaction to Jerry after I regained my composure and some color returned to my face was that he had made an awful mistake. Nothing seemed to add up. I asked Jerry why *he* was delivering the message. According to him, as the original producer of *Cats* he'd struggled raising the money and had to turn to Mr. Gennaro for funding. Jerry owed the Mafia and thus did small favors in return. I'm not a theater expert, but I did know that *Cats* was created by Andrew Lloyd Webber, who is the most successful Broadway composer of all time. He had the track record to raise millions of dollars from legitimate investors. Why would he need the Mafia to fund his musical?

Sensing I doubted his story, Jerry pulled out his cellphone and dialed a number. Handing me the phone, the voice on the other line introduced himself as Mr. Gennaro, no first name. Jerry was so disgusting I held the phone away from my mouth and ears afraid I would contract a disease. He retold Jerry's version identically, including the threat. I blew four million dollars of his money on a bad Yahoo! stock tip and unless I paid him back by tomorrow, I would be killed. He then abruptly hung up. The entire call lasted less than two minutes. I should have been terrified, except Mr. Gennaro's voice sounded like a very bad *impression* of a New York Italian Mafia Don. The entire experience seemed unbelievable. I was waiting for the terrible joke to end.

As I stood to leave, Jerry warned me that people were watching us and I'd only be safe walking out with him. I wanted to run as far away from him as possible, but I agreed to let him walk me to my car. Once in the parking lot, my eyes became telescopic. I was acutely locked onto everything around me trying to identify if someone was fixed on me. On the way out, Jerry walked me first to his car. There was one last item I had to know about. When he opened the door to his car, I peaked in. It looked like it had just come from a dump. Inside were McDonald's wrappers all over the floor, soda and beer cans tossed on the seats, and, almost amusingly, both boxes of condoms and right on the driver's seat, a bottle of Viagra. Jerry, not sensing I was looking inside his car, pulled out documents to prove he was who he claimed to be. They included a movie script with signed agreements from Al Pacino and Dustin Hoffman as the leads. The contracts looked valid, but after all I'd seen and heard over the past thirty minutes I didn't even know if *my* existence was real, let alone signed documents from major Hollywood stars.

Inside my car, I was about to pull away when a huge, terrifying looking man walked over to my window. He just stood above me and stared at me for about thirty seconds. In the past twenty minutes, I had been petrified, confused, amused, and then petrified again. The man never said a word. He just walked away. I turned on the ignition and hauled ass out of the parking lot. On the way home, I made several turns to make sure I wasn't being trailed before I was comfortable I was alone.

Once home, I tried to collect my thoughts and make rational and reasonable decisions on my next step. There was no way I could tell Sharon what had just occurred. I needed a plan first. On the disbelief side, there was this drunk, seemingly insane, weak-looking man trying to convince me he was a major Hollywood producer working for the Mafia. There was a phone call with a man with an obviously fake Italian accent. On the believable side, there was a monster of a man who scared the hell out of me. Jerry had warned me against calling the police. Not knowing what I was up against, I left calling the sheriff's office out of it for the time being.

Replaying every word from our discussion, I recalled Jerry claiming that Mr. Gennaro and I had met two years earlier at the Marriot Marquis. The only time I remembered being at that hotel was two years earlier was when I met Mark Geddis, my old friend and future business partner, which was in the middle of the day. I was convinced I hadn't been there for dinner or drinks when we supposedly had met.

For two years, I couldn't get Yahoo! out of my mind. It consumed my thoughts. I'm sure that helped contribute to Convizion's lack of success. I'd daydream and place myself back there in 1999. Often however, I wished my tenure at Yahoo! had never occurred. In the past two years, it had caused lots of trouble in my life. At the age of thirty I had no plan for the rest of my life. I lost all confidence in recapturing the excitement of my twenties. This brought on depression and an early midlife crisis and now, I had a Mafia chief claiming that he lost four million dollars because of my Yahoo! stock recommendation. I could have just remained in the advertising industry; by then, I'd have a salary of close to one hundred thousand and no depression or drama.

When I was through cursing Yahoo! and my life, I regained my composure and decided I needed some help. I called my attorney, Kevin McBride, and asked him to come to my house. I told him I was afraid to go to his office, that I was going to be tailed and that my phone could possibly be tapped. Kevin was definitely alarmed, but nevertheless didn't push and rushed to my house. I meticulously replayed for him the entire Jerry ordeal. His suggestion was that I call a private investigator and he had just the right person in mind.

Wayne Black was a former Miami detective during the 1980s Miami cocaine wars. After leaving the department, he worked as a senior executive for a private security company before launching his own private investigative firm. Wayne had already gained national prominence. He was contracted by Exxon after the Valdez oil spill in Alaska to help minimize the fallout resulting from congressional hearings. He had been featured on *60 Minutes* and several morning news programs. All the major national newspapers prominently wrote about him. He was well known for helping families track missing children and protecting celebrities from dangerous stalkers.

Kevin convinced me to go to his office to meet Wayne. When Wayne arrived, he was a tall, older, good-looking gentleman. He had a full head of gray hair with a gray goatee. He dressed casually and looked as if he had just been pulled from a boat off the coast of Key Largo. He had a huge smile which looked as if were painted on permanently, and he was very welcoming to Kevin and me.

After some pleasantries and background exchanges, in great detail and illustration I replayed the entire conversation with Jerry Davis from earlier in the day. I spoke of the beat-up Town Car, the garbage spread all over the seats and floor, including the condoms and Viagra, Jerry's appearance, Mr. Gennaro, and that horrible Italian accent.

Wayne listened intently and seemed genuinely perplexed. For someone who had worked thousands of low- and high-profile cases, he'd never heard a story this odd. Right away, he knew the Mafia wasn't involved. In his years of experience chasing Mafia figures, he'd never witnessed any of them driving cars that were anything less than impeccably clean. In addition, based on my account, there was no way Jerry had produced *Cats* or had any Hollywood connections. At first I thought that was good news, but according to Wayne, it wasn't. If the Mafia really was extorting me, Wayne could soothe or diffuse the situation. Wayne knew the names of all bad guys in Florida, but he'd never heard of Jerry Davis. Coupled with such an outlandish story, he was concerned that he could be dangerous, unpredictable, and possibly part of a larger network of extortionists.

It was strongly suggested that Sharon, our two young boys, and I find a safe place to temporarily relocate and that we should hire bodyguards to watch us 24/7. Wayne reasoned that Jerry already knew where we lived and would likely be following us.

I was instructed to go home and pack up. Wayne and Kevin escorted me back home to ensure my safety. I knew Wayne was well-armed and surprisingly, he was very calming. In his long career in public law enforcement and private sector criminal justice, there's not much to which he hadn't been exposed. I felt safe.

I had a strong suspicion that explaining to Sharon that she had to leave her house and travel with a bodyguard because a mad-man was demanding four million dollars or he'd kill us all was not going to be a pleasant conversation.

Not surprisingly, I was right! Dropped jaw, silence, and catatonic were her reactions. After several minutes, her color slowly reappeared, her mouth moved, and words actually became more coherent and she was able to lower her hands, which moments earlier had been stuck to her cheeks. I recounted the entire episodes from the day, including a blow-by-blow account of my meeting with Jerry. Wayne assured her that this was a scam and not the Mafia trying to kill us, but Jerry Davis was unpredictable and had to be considered dangerous. Sharon's response was revealing. Rather than blame me for meeting with Jerry or blame Wayne for her being uprooted, or even blame Jerry himself for tearing our life apart, she blamed Yahoo! If it weren't for Yahoo!, she reasoned, we wouldn't have enough money to be scammed. Also, if it weren't for Yahoo!, we'd be anonymous, which at that moment both of us desperately prayed to be.

I was blaming Jerry, but I understood her anger toward Yahoo! This point was the culmination of two horrible years since I'd resigned and we'd moved to Florida. There was definitely a part of me that wished for an average job with an average home and an average lifestyle.

That night, Wayne stayed at our home to make sure we were safe. I couldn't sleep. I continued to wonder how my life had gotten to this point. I was caught in this surreal moment and didn't have a way out. All I knew was that the next few days were going to be unlike anything I could ever imagine.

The next morning, Wayne arranged for two bodyguards, Gil and Michael, to accompany us, one for Sharon and the boys and one for me. Gil and Michael appeared at our home at daybreak and we were introduced to them.

Gil was older, probably in his late fifties. He was of Cuban descent, short, a tad overweight, and balding. He had been a former undercover Miami detective who'd worked alongside Wayne during the 1980s cocaine wars, the era that was memorably illustrated in the movie *Scarface*. Gil had stories of stings against Central American drug lords importing cocaine into this country. I knew just how real he was when he recounted the time he was taken hostage by a Jamaican drug cartel. About to be assassinated, the gun malfunctioned, and fortunately for Gil, the police SWAT team stormed in to rescue him.

Michael was of Austrian descent. His accent was identical to that of Arnold Schwarzenegger. He was short, well built, blond-haired, and probably in his mid-thirties. Michael was intimidating. He spoke without a touch of personality, direct and in short compact sentences in his thick Austrian accent. He carried a bag over his shoulders that scared me just considering its content of heavy firepower. Gil was going to guard me, and Michael was assigned to guard Sharon.

The first order of business was to decide where we'd stay. I decided that if we'd have to go into hiding, we should do it in luxury. I suggested the Boca Raton Resort and Spa, a five-star resort on the beach in Boca Raton, Florida. No argument from Sharon. Wayne happened to be friends with the head of security at the hotel and he agreed we'd be safe there. He arranged for our room under an alias.

Once we checked into the hotel, I was given a small recorder and instructed to tape any conversations I had with Jerry. On cue, Jerry called with the instructions for the drop-off point. Wayne's plan was not to meet Jerry immediately. He believed if we made Jerry wait a few days he could flesh him out more and build a more solid case to give the police. I told Jerry that I needed a few days to get that kind of money together, that I needed to sell stocks in order to raise the capital he was demanding. Jerry continued in character by letting me know he'd need permission from Mr. Gennaro and he'd get back to us. About ten minutes later, he was on the line again. When he called back his tone was vicious. Although the request to wait a few days was granted, with emphasis, he warned if I screwed them I was a dead man. He added that he knew I had left our house and the Mafia was trailing us at all times. I assumed he had already calculated we had deserted our home but I wasn't convinced he knew where we were hiding out. I recorded the call, and we attempted to trace the call, but it was useless. He called from a throwaway cellphone.

We abandoned the house so quickly that we forgot several items we needed for the baby. We had two children, one had just turned two and the other was just three months old. Both needed basic necessities such as formula and diapers. Michael agreed to go to Target, but Sharon, already feeling claustrophobic, demanded that she go along. Also, she didn't trust Michael to get the right brands. Michael was a mercenary who had once been hired to guard Hamid Karzai, the president of Afghanistan, and other world leaders who lived in daily fear of being assassinated.

I was also getting stir-crazy so I decided to go as well. We parked at Target and Michael instructed us to stay in the car so he could case the lot. He looked inside and underneath parked cars. He eyed everyone walking into the store and the parking lot and finally returned to the car. In his thick, heavy Austrian accent he announced that the "perimeter is secure." I found it ironic how this trained killer who was used to guarding leaders of Third World nations during civil wars and revolutions was securing the perimeter for us at Target.

After the death-defying trip to Target, we returned back to the hotel. That evening, Jerry called again to remind me that I needed to secure the money. He also declared that the Mafia was now targeting Sharon and the kids. We recorded the entire call. For obvious reasons, Wayne suggested keeping this new information from Sharon.

The next day, Wayne suggested we go to the office to speak to Corrigan and the others who were in the office the day Jerry appeared. Gil escorted me in his car. Aside from enjoying some time with friends we learned nothing new. The guys were all happy to see me, but couldn't believe what was going down. They had already changed the locks and set up a new alarm system at the office.

After leaving the office, Gil and I decided to get pizza at a local restaurant called Sal's Pizzeria. Once we were comfortably back at the hotel, Jerry called again. This particular conversation alarmed Wayne. The anger in his voice suggested he was a crazed lunatic. I could almost feel the saliva from his mouth through the phone as he insanely and, at times, incoherently screamed. He demanded to know who the short bald man was who'd accompanied me to Sal's. Obviously, he was describing Gil. Gil was trained in identifying when a victim was being followed. He was confident that he checked his surroundings when we entered and left the restaurant and recalled nothing suspicious. I knew exactly what Jerry's pathetic rusted eyesore of a car looked like. I didn't see it, and I was on the lookout for it.

Wayne was concerned. He strongly suspected Jerry was working with a larger network. I was already scared, but now I was petrified. Wayne was always calm, so when he was nervous, it almost threw me over the edge. I likened it to flying. I only worry if the pilots worry. Wayne was the pilot, and he was worried.

Jerry had demonstrated his temper before, but this was unlike any hollering I'd ever heard. He reminded me his orders were not to seek any police or professional help. Ironically, he said he could no longer trust me. The Mafia now demanded six million. According to Jerry, my fuck-up had just cost me two million dollars.

Wayne was upset that we still hadn't been able to identify Jerry. The fact he sounded like a deranged individual with the resources to have a team supporting him made the ordeal more dangerous than we had initially suspected.

Sharon and I are big sushi eaters. We order it every Sunday night. While the hotel had some wonderful restaurants, we craved sushi and convinced

Wayne to allow us to go out to eat. We chose our favorite Japanese restaurant and with Michael and Gil joining us had a very pleasant dinner. I was in a better mood. It was two days after we were busted eating pizza and we hadn't heard from Jerry. Although the deadline for paying the six million was only forty-eight hours away, I was cautiously optimistic he had been scared off when he realized we were not alone.

Unfortunately, my relief lasted only until the next morning when I received another terrifying call from Jerry. Unlike our last conversation, he was calm. He asked how the sushi from the other night tasted. The question caught me off-guard. He was still following us. It was the only time in two days we'd left the hotel and the psycho was trailing us. Again, Michael and Gil were both stumped. Neither had noticed anything or anybody out of the ordinary the entire night. Jerry reminded me that he warned me to be alone, and I wasn't. He demanded to know the identities of the other men at the table. I tried to convince him they were friends, but he refused to believe it. I'd defied his orders and now I had to know just how serious he was. According to Jerry, he paid a busboy to poison my sushi and in exchange for the six million he would give me the antidote to save me. He reminded me of our meeting the following day, and then he hung up.

As usual, I recorded the conversation. When Wayne listened to the tape, he burst out in laughter. In all his years in law enforcement, he'd never heard a threat so pathetic. Although it amused him briefly, he was still troubled. While no one seriously thought I'd been poisoned, we still had grave concerns about Jerry's stability and his ability to finance a team for the extortion. Someone *was* following us and it was an individual or a group that trained experts couldn't identify.

Jerry called again with instructions where to meet the next day. Wayne was adamant that we meet in a public restaurant. We agreed to meet at the

Sunset Diner, which was across the street from my office. Jerry warned me to be alone.

Traditionally, I am a very good sleeper. As soon as my head hits the pillow, I'm out. Stress has never kept me awake. Once I'm asleep, I have the ability to sleep peacefully until morning. That night, I stayed awake. I imagined every scenario possible. I visualized being kidnapped and even being shot. Wayne and I had rehearsed the entire scenario. The diner was like an old rail car. It was narrow with booths on either side, with barely enough room for the servers to walk the single aisle down the middle. I had only eaten there a few times and especially enjoyed their vanilla milkshakes, for which I always had a weakness.

The plan Wayne devised required me to excuse myself to the bathroom when Jerry appeared. Wayne, Michael, and Gil would surround Jerry and quietly warn him they were capable of shooting him if he tried to escape. They were going to interrogate him and identify the entire plot. They planned to allow him to leave, trail him, and finally call their connections at the local sheriff's office to arrest Jerry at a later time.

The next day at noon, I arrived at the diner. Wayne had contracted with six other private investigators, all strategically sitting at different tables throughout the restaurant. I sat down at a table in the middle of the restaurant facing the entrance. I was visibly shaking. I had my guard up the entire time and was very methodically checking each person to determine if I was being watched. No one looked obvious to me.

Jerry was scheduled to arrive at twelve-thirty. I'd arrived early and to help calm my shaking nerves, and I ordered a cheeseburger with French fries and a vanilla milkshake. At twelve-thirty, Jerry hadn't yet arrived, one o'clock passed, two o'clock passed, and finally three o'clock passed. Still no sign of Jerry.

Unfortunately, but not unexpectedly, he never appeared. After three hours, six milkshakes, and a life-planning session, Wayne passed a note instructing me to leave. Jerry was supposed to be at the diner at twelve-thirty and it was already past three o'clock. Gil was waiting outside to drive me back to the hotel. Just as I walked out the door my phone rang and on cue, Jerry was on the other line demanding I go back in the diner. I was flabbergasted. He clearly had an unidentified spy in the restaurant. Wayne and his colleagues had watched every booth the entire time and witnessed nothing out of the ordinary. It was apparent Jerry either had paid someone who was working at the restaurant or strategically positioned people coming in and out throughout the afternoon.

My newfound resolve allowed me to tell Jerry to f-off. I emphatically declared that I had waited for three hours and I wasn't waiting another second. I couldn't believe the words came from my mouth. As I ran as fast as possible to Gil's car, Jerry made a fatal mistake. Cursing Jerry caused him to snap. He forgot my name and referred to me as "Joe Mac."

Joe (Mac) McGuire was a sales rep we had hired in New York for Convizion six months earlier. Joe was a decent employee. He was quiet, never caused any problems, and was well below the radar in our organization.

I jumped in Gil's car with a wide smile. We finally had a lead. Rather than returning to the hotel, we convened at Wayne's office. I called Joe, placed him on speaker so everyone could hear and described the entire experience of the past several days. I detailed Jerry's features: no teeth, rancid, rail thin, heavy smoker, and so forth. Joe instantly identified the perpetrator. Jerry Davis was an alias. His real name was Tom Bennett.

Joe not only filled in the answers to several outstanding questions, but provided tremendous insight into Bennett's recent past. A year earlier, Jerry/Tom had successfully scammed Jamie Brophy, a colleague of Joe's in Arizona. Jamie owned a successful business that marketed children's video content for feature films, documentaries, and television programs. Joe worked as an East Coast sales representative for her company prior to joining Convizion.

Bennett stole over one hundred thousand dollars from Jamie in a scam. He befriended her, developed a close relationship with the entire Brophy family, and ultimately established a business partnership. He convinced Jamie to invest her money in a "failure-proof" project and ran off with her life savings. We were amazed that she had a business arrangement with Bennett, but at that moment, our primary concern was locating him.

Jamie had introduced Joe to Tom Bennett who had no indication that Bennett was a bad guy. They'd never met face-to-face. Joe only knew that Bennett and Jamie had a business arrangement. New to Convizion, Joe rightly assumed he'd make a good first impression on my business partner Mark and I if he could close a sale in his first week. He reached out to Bennett with a proposal to sell stock footage to use in his upcoming movie.

Hoping to impress Bennett, Joe pitched the experience and background of the Convizion executive team, including their young CEO, Jeremy Ring, a former early Yahoo! executive who had made a killing on its stock. Bingo! All the bells in Wayne's office went off at once.

I explained to Joe another man was involved. I described the awful Italian accent. Joe assumed it had to have been Spatz Jordan, a close associate of Bennett's. Spatz was an out-of-work actor who had always dreamed of doing voiceover work but who had never succeeded. Although he never personally met Spatz, he'd spoken on the phone with him several times. Joe was confident that Spatz was Mr. Gennaro. Wayne believed the money Bennett stole from Jamie was used to recruit other individuals for the extortion plot to target my family and me.

Joe recommended we interview Jamie to gather more details and then warned us that he was terrified of Bennett, whom he believed was a very dangerous individual.

We reached Jamie at home. After introducing ourselves there was a long, silent pause. Then, without warning, she broke down so uncontrollably that her words were incoherent. Unable to gain any composure, she had to hang up. Before ending the conversation, Wayne gave her his number and requested to speak to her again during a less stressful time, to which she agreed.

Locating Tom Bennett was virtually impossible. His phone was a throwaway, and we didn't have the registration for the Lincoln Town Car, therefore making it impossible to track the owner's whereabouts. With such a common name as "Bennett," it was impossible to identify which Tom Bennett was extorting millions from me.

Wayne was proud of me for telling Bennett to f-off, but his concern continued. We still had to assume he was dangerous and that he wasn't working alone.

That night Michael stayed outside our hotel room until daybreak, which made me feel safe. If there was any attempt to harm us, Michael would shoot them before they could blink.

The next morning we met again at Wayne's office. Fortunately, Jamie had called him back the night before. She was vehement that she would speak to Wayne only once and then would have no further communication with us. She warned Wayne that Bennett was very dangerous and she believed he had killed before. He carried weapons in his car, and he'd explained that he did so because the Mafia was chasing him and he had to protect himself. Jamie, through her tears, described how Bennett had ruined her life. She trusted him and he stole one hundred thousand dollars, which was her life savings. She had to close her business and was now struggling emotionally and financially every day. He had damaged her profoundly. The positive news was Jamie was able to tell us that Bennett lived in Orlando, Florida, and had a daughter named Tina. With that the call ended, and that was the last we heard from Jamie.

Ironically enough, Bennett, who usually checked in daily, didn't call us that day. Wayne was able to obtain an address and the next day drove to Orlando to confront Bennett personally. He arrived at a trailer park, that, according to Wayne, was quite similar in stench and filth to the Town Car, just on a larger scale. He knocked on the door and Tina answered. Tina was very suspicious, but nevertheless invited Wayne into the trailer. The smell was so awful that Wayne's eyes were tearing. Clothes and rotten food littered the floors and along with Tina was her mother, Tom Bennett's estranged wife. Tina was a petite redhead, in her early thirties who, if she wore make-up and had her hair styled, could have been attractive. She had a large tattoo showing from her right breast and her teeth were crooked and looked as if they hadn't been brushed in years. She was a heavy smoker with a terrible cigarette odor emanating from all over her body. Tina's mother was extremely overweight, according to Wayne probably close to four hundred pounds, and likely hadn't showered in weeks.

Wayne introduced himself and then openly and honestly described the entire saga. Both Tina and her mother listened carefully, but believed none of it. According to Tina, Tom was an outstanding man. He was a successful man. Her father was a big name in the entertainment industry. She proudly boasted to Wayne that her father had produced the hit Broadway show *Cats* and was about to produce a major Hollywood movie with Dustin Hoffman and Al Pacino. However, for all of his success, she warned that he had serious friends, men who had committed terrible crimes and were very dangerous. According to Tina, her father was associated with the Mafia.

Wayne was stunned. Tina Bennett and her mother were living a life that was beneath even Jerry Springer standards and had been convinced by Tom that he actually was everything that he portrayed himself to be to us. The most troubling thought was not that he lied to his family, but that they believed him. He was a complete degenerate. His appearance was even more horrifying than the trailer home, and that was disgusting.

We were disappointed to learn they hadn't seen him in weeks. They believed he was in Los Angeles, filming a future Academy Award-winning movie. Sternly, Wayne instructed them to call him if he appeared. Just as he was leaving, Tom's son, who lived in the trailer next door, stormed inside in a fit of rage, declaring that his father had just stolen his gun. With that, Wayne left and drove back to South Florida.

On his way back, Wayne, troubled that Bennett had stolen a gun, reached out to the Broward County sheriff's office. He'd compiled enough information to turn the case over to the authorities.

The following day, Wayne and I met with Detective Sharp. Armed with the knowledge of having identified Tom Bennett, along with providing his address and the background information from Joe and Jamie, we described the entire ordeal. With official permission granted to record my conversations with Tom we were instructed to arrange a meeting in hopes the police could arrest him if he actually appeared. Although we were recording all the prior conversations, we did so without a court order and therefore they would not have been admissible in a trial.

Later that day Jerry/Tom called. It was clear he hadn't spoken to his family because there was no mention of Wayne's meeting with Tina and her mother. By this time, I had gotten to know Tom well enough that if he knew about the meeting he would have thrown a fit. However, he likely realized that I recruited help. In a semi-angry tone, for him, he declared he was sick of all the games I was playing and would no longer meet me. His new tactic was to demand that I wire all the money to an offshore account. In my newfound empowerment, I rejected a wire transfer. I held my ground and announced to him if he wanted the money he would have to meet me face-to-face. Instantly, Tom got agitated, again threatened my life, and slammed the phone down.

Ten minutes later he called back and, in a very calm tone, agreed that he would meet me. I set the meeting for lunch the next day at Houston's, a popular family restaurant that was sure to have a crowd. I reminded Tom that the last time we had agreed to meet, he had not showed up, and I wasn't going to wait all afternoon for him. Tom, in typical fashion, grew angry and reminded me that I walked out of the diner before he could arrive. After we hung up, I was angry with myself for forgetting to inform him that the poison sushi hadn't yet killed me.

The next morning, I arrived at a satellite police station. Detective Sharp and other Broward County sheriff officers greeted me. Once there, a wire was taped to my chest and a microphone was hidden in my pocket. While being taped up, I engaged in a discussion with the other officers. They were all members of an elite SWAT team who were trained, like Michael and Gil, to shoot first. They proudly declared if they shot, they wouldn't miss and warned me that if guns were pulled, I should take cover under the table. I didn't need the reminder. I had practiced it in my mind all night.

Unlike the day at the Sunset diner a week earlier when I'd arrived petrified, I was excited and ready. I was hoping that Bennett would appear. Adrenaline rushed through me. It was thrilling and I was pumped up. Tom Bennett no longer scared me. I felt exhilarated.

The restaurant was packed, and at least three tables were filled with SWAT members, ready to shoot Bennett right between his eyes if he slipped up even the slightest. In my mind, I couldn't believe where I was in life. No one could make this stuff up. I was placing together a timeline in my head so I could one day document this entire remarkable episode. So much had happened in the past week that I had to struggle to recall all of the events and the order in which they occurred. I was only two years removed from Yahoo!, yet it seemed like a lifetime ago.

After two hours, I was still alone at my booth and I knew Bennett had stood me up again. I wasn't surprised, but I was disappointed. Detective Sharp motioned me to leave. Gil was again waiting outside for me. As I walked to Gil's car, two men jumped into their car, a red Chevy Blazer and attempted to follow us out of the parking lot. Gil and I noticed it immediately. Prior to pulling out of the parking lot, three unmarked police cars swarmed the men, dragged them out of the red Blazer, pulled them to the ground, handcuffed them, and drove them to station.

Neither man was Tom Bennett or Spatz Jordan. Interestingly, unlike Tom or Spatz, both men were clean-cut and looked harmless. For over an hour Detective Sharp interrogated them. They demanded to know why they jumped in the Blazer so quickly, why they chased Gil's car, and what was their association with Tom Bennett? Interestingly, they each had an airtight alibi. Both men claimed they were out-of-towners on their way to a golf outing with two business associates whom they had never personally met. In their version, the plan was to locate a black Chevy Tahoe, which describes the vehicle Gil drove, in the Houston's parking lot, and follow the car to the course. According to them, they mistakenly identified Gil and me as their golfing partners, thus as Gil drove away, it was their cue to follow.

Neither Detective Sharp nor Wayne believed one word, but their alibis checked out. The police contacted the supposed, never-seen golfing partners who validated the story. Unfortunately, the two men were released without any charges. Tom Bennett never called again.

Positive that the two men in the Houston's parking lot were working for Bennett, we assumed they warned him his scam was up. Detective Sharp notified the Orange County sheriff's office, Bennett's home county, to be on the lookout for an old, beat-up blue Lincoln Town Car with fast-food wrappers, condoms, and Viagra in the front seat. Just in case, Wayne called

Tina and described the events of the day, hoping she was in communication with her father.

Wayne was confident that Bennett wouldn't attempt to contact us again, but to be safe, suggested we not yet go home. I was angry that Bennett was still on the loose, and felt a tremendous resolve to see him held accountable. Although I knew the financial costs of using Wayne, Michael, and Gil was growing exponentially, I nevertheless instructed Wayne to hire a private investigator in Orlando to locate Bennett. I have tremendous respect for law enforcement, but I doubted they were going to dedicate someone exclusively to finding Bennett and that's what I wanted. For eight days Bennett had chased me and made my life hell. If it took me the rest of my life, I was going to chase him and make his life a living hell. Part of me was angry, but another side of me was exhilarated. I had the resources to go after Bennett and I was going to make sure he was captured.

Fortunately, it didn't take a lifetime; it took less than a week. I was a pest to Wayne. I called him several times a day for an update. For five days, there was no news, but on the sixth day, Wayne called to deliver the news that the investigator we'd hired in Orlando had located Bennett. He and Spatz were sleeping in his car in a Publix shopping center parking lot. Wayne instructed his partner to wait until he arrived before calling the police. He wanted to be there to give the sheriff's office his report directly. He jumped into his car drove three hours to Orlando. Once there, they contacted the Orange County police who swarmed the Lincoln, dragged Bennett and Spatz out of the car, and arrested them. Wayne reported that Bennett was screaming and crying like a baby while Spatz was trying to hold it together, unaware why he was being arrested. Pictures of the car were taken and emailed to me. The wrappers, condoms, and Viagra were still in the front seat. I was so amped up. I was angry at myself for not witnessing it all. I would have probably stomped on Bennett right in front of the police without any care of the consequences.

The car was so filthy the police had to wear masks when searching it. One even remarked to Wayne, he'd never seen anything as disgusting as the inside of that car.

The search turned out frightening evidence. In the trunk were Dustin Hoffman's and Al Pacino's signed contracts and a tape recording of Spatz practicing his awful Italian accent. Alarmingly however, there were pictures of me and a movie script titled *The Great Kidnap*. The script was a story of scaring an internet multimillionaire into believing he had met a Mafia head by the name of Mr. Gennaro in New York and had given him a stock tip on his company, which led to the loss of millions of dollars. The Mafia poisoned his sushi and had the only antidote that could save him. The story continued that had the main character not paid or not believed he was poisoned, he was going to be taken hostage to force the family to pay.

In the script, the kidnapping was executed at gunpoint by two men wearing motorcycle helmets jumping out of their car and duct-taping the internet millionaire's mouth shut. Dustin Hoffman was cast as the internet millionaire and Al Pacino was cast as the kidnapper. Hoffman's character was named Jeremy Ring, and Pacino's character was named Jerry Davis.

Also recovered in the trunk were duct tape, two motorcycle helmets, and a 9mm handgun, which we discovered later was registered to Bennett's son.

Bennett was so deranged. His *real* plan was to execute everything in the script and then sell the movie to a major Hollywood studio with Dustin Hoffman and Al Pacino as the leads. Below the title of the script *The Great Kidnap* was the phrase "Based on a True Story."

Bennett was ultimately extradited back to South Florida, where not surprisingly, he skipped bail. He was captured in a North Carolina Motel 6 one year later. With his bail revoked, he was sent to jail while awaiting his hearing. He invoked an insanity defense, and did his best impersonation of acting crazy during the short trial. Fortunately, the judge wasn't fooled.

Bennett was found guilty of one charge of extortion and was sentenced to five years in the state prison with ten years of additional probation. He was incarcerated in a prison hospital, convincing the prison officials that he was an invalid. He was released in 2009 and to date has never produced his Destin Hoffman/Al Pacino-starring movie. And I'm still alive, even after ingesting the poisoned sushi. Every so often we check up on him, wishing he'd break his probation in the hopes he'll be sent back to prison for the rest of his sad, pathetic life.

With this entire ordeal behind me, I set forth to execute the plan I devised that day at the Sunset Diner. I was reborn. I had a newfound excitement for life. I was going to create something that would eclipse even Yahoo! There was just one major challenge I hadn't anticipated—that that would be really hard to do.

CHAPTER 8
Google: The Largest Missed Opportunity in the History of Business—Take 1

On July 25, 2016, Verizon Communications announced it had acquired the operating businesses of Yahoo! for 4.8 billion dollars and would be merging Yahoo!'s assets with those of America Online (AOL) to form a new company, inappropriately named "Oath."

It's as if the leaders of the two companies took an "Oath" to merge two sinking ships. When they are officially declared dead at sea, the one-time adversaries would be partners at the bottom of the ocean for eternity.

Numerous questions have been asked, and will continue to be asked for years, that we will ponder and study. How did Yahoo!, one of the great brands in corporate history, in only twenty-two years, rise to some of the greatest heights ever seen in American business and then fall into oblivion?

For anyone paying attention, the beginning of the end began with decisions made by the first team while the company was on its way up. Some actually set the stage for the horrific decisions made by subsequent generations of Yahoo! leadership. Most were driven either from pure incompetence or just lack of vision by various CEOs from 2001 to the present.

Employees, reporters, users, shareholders, and observers can debate forever why Yahoo! failed. Most have a short memory and place all the blame on Marissa Mayer, the last CEO of the company. While she deserves some fault, I'm convinced that had Steve Jobs, Jack Welch, Lou Gerstner, Bill Gates, George Washington, or Abraham Lincoln become CEO of Yahoo! on July 16, 2012, none of them could have reversed the fortunes of Yahoo! Poor decision-making, arrogance, lack of vision, and poor execution prior to Mayer starting had already cemented the legacy of failure.

The Yahoo! story is a story of missed strategy, execution, and opportunities. Early decisions to de-emphasize search, to not acquire Google, and to ineffectively manage the tension between the product teams responsible for user experience and the sales team responsible for advertising experience created the foundation for the company's future downfall. Also, factors outside of Yahoo!'s control—most notably unreasonable expectations of Wall Street analysts, mutual funds, day traders, and hedge fund managers—created an environment where short-term decisions were made at the expense of the long-term good.

Decisions and poor executions by the "second" generation of Yahoo! leaders resulted in a choking bureaucracy led by a CEO with the wrong kind of "deal-making" experience for an internet company. The brashness and arrogance of executives to overplay their hand when they could have, and should have, acquired Facebook, has haunted the rest of their existence.

The subsequent CEOs (prior to Marissa Mayer) arrived so late to the emerging mobile platform that Yahoo! was nearly irrelevant as users moved from their desktops to phones to manage their daily habits.

Losing search, video, and mobile services to Alphabet (the parent company of Google, YouTube, and Android), and social and communications to Facebook all occurred *before* Marissa Mayer started. There is no possible way to overcome a Grand Canyon-sized deficit, without either unlimited capital to acquire fast-growing upstarts (such as Instagram, WhatsApp, Twitter, Snapchat, and so on) or having a clear understanding of a corporate strategy, neither of which Yahoo! afforded.

Twenty-one years after its incorporation and sixteen years after its stock peak, Yahoo! sold for 96 percent *less* than its value on January 3, 2000, when it closed at an all-time high of 118.75 (split-adjusted) dollars per share, resulting in a market capitalization of 120 billion dollars. On that day, Wall Street valued Yahoo!, in business less than six years, higher than Disney, News Corporation, and Comcast combined, and even higher than Ford, General Motors, and Chrysler combined.

At the end of 2016, the top seven businesses on the list of the highest-valued companies in the world by market capitalization included Apple at #1, Alphabet (Google's parent company) at #2, Amazon.com at #5, and Facebook at #7.

On January 3, 2000, the iPhone was still more than seven years away from launching. Apple was, at best, a floundering company, that actually had been saved from oblivion, through an investment from their arch-nemesis Microsoft two years earlier. Google was four years from its Initial Public Offering (IPO), Amazon was hemorrhaging money, and Mark Zuckerberg, Facebook's founder, was still in high school.

Most of the civilized humans on planet Earth have, at one time or another, complained that their lives did not turn out the way they had hoped: the high school star athlete working paycheck to paycheck, the local rock-star desperately trying and failing to hit it big on the charts, the talented writer now working as an overnight security guard, and the Yahoo! leaders having the opportunity to acquire Google, eBay, and Facebook and either blundering it or turning it down.

Keep in mind that Yahoo! sold its entire operating business for 4.8 billion dollars. The combined market capitalization of the three companies listed previously in July 2017 is over 1.2 trillion dollars. While never confirmed, it's possible Yahoo! had informal internal or external discussions to acquire Apple, Amazon, and Netflix, when each of those companies were either flailing

(Apple) or in its infancy (Amazon and Netflix). Taken into consideration those organizations have a July 2017 combined market capitalization greater that 1.35 trillion dollars. (Because of its market position, it's likely Yahoo! would have had exposure to all the major technology deals.) Kind of makes you want to kick yourself just a little.

In 1998, Larry Page and Sergey Brin, Google's co-founders, offered to sell their PageRank system, which was the core of what would become the Google Search Engine, for one million dollars. In July 2017, Google's market capitalization was valued at over 675 billion dollars. With fairly strong conviction, I'll go out on a limb and flatly state that Google's one-million-dollar price tag in 1998 was probably the best deal offered in the history of Silicon Valley, California, the United States, planet Earth, and the Milky Way Galaxy. My guess is most individuals are probably kicking themselves for missing that one.

The real reason the Google deal was rejected was that in 1998, Yahoo! had grand plans to move away from being a search engine. It has been well established that Jerry's and David's original intent was for millions of people to better navigate through a maze of untenable amounts of information. If Netscape was the front door of the digital house, then Yahoo! was its main foyer. Its purpose was to send people to different rooms to find whatever their hearts desired. There could be a music room, a sports room, an entertainment room, and even a trampoline room. There was something on the internet for everyone, and the role of Yahoo! was to guide people on their way. It was, for all intents and purposes, the main check-in counter. Beginning in 1997, Yahoo! lost the very soul of its initial purpose. Yahoo! decided they never wanted their customers to leave their party. Although it was nineteen years of a slow drip to death, a strong position can be argued that persuading users to stay rather than to leave was the first sign of poor strategy that haunted the company for the rest of its existence.

Around this time, Yahoo! was also adding content furiously. They created a website for kids, launched classic games, and partnered with content providers to create news, sports, entertainment, and finance products. They allowed users to book travel, map directions, shop for products, find companionship through personals, and sex through its chat service. A user could go to Yahoo! for a job search, post classifieds, see their daily horoscope, and read the day's comic pages from newspapers all over the world.

Yahoo.com had evolved into the Yahoo! network of sites. As it did, the search feature was becoming less relevant. At one point, the company even discussed and designed mock-ups of the home page that had a search bar at the *bottom* of the page where a user would have to scroll down to find it.

By 1999, being a search engine was already out of favor and being a "portal" was the rage. Keep users on Yahoo! for all of their needs. Be the one place to go for all of your online needs—be all things to all people. But that strategy was fraught with unforeseen peril. There are over one billion websites across the globe. In 1999, the number was much less, but still clearly growing exponentially. The universe was expanding to limitless heights. Therefore, more than ever, the greatest need was for a jumping off point, i.e., Yahoo!'s original purpose. There were thousands of websites that grew to millions of sites that offered news, games, entertainment, commerce, and so on. Almost all of the content on the Yahoo! network was ubiquitous across the entire internet so there was no possible way to consolidate its users. It would be like every broadcast and cable network being consolidated into a single channel. It's about as far from practical as a trip to Jupiter would be. While Yahoo! was growing its user base to over one billion strong, the company incorrectly believed that those billion users would primarily use services found within the Yahoo! network.

To be sure, there was an interest in Yahoo!'s services. Individuals across the world used the network to find stock quotes, sports scores, entertainment news, and other information that aligned with their personal content fix. What was lost on the Yahoo! product group, though, was the realization that the content offered to users lacked depth or exclusivity. Users visited NewYorkTimes.com for detailed news content, they visited Ticketmaster or a music fan page to learn more about bands and their tours, a school paper was researched on academic websites, and so on. In fact, the most pervasive use of the internet, its darkest secret, is that the words most searched for weren't Brittany Spears, Pearl Jam, or the New England Patriots, but rather for sex and every possible word sick minds can ever think of related to sex. What this meant was millions of individuals were using Yahoo! primarily to leave Yahoo! For millions, their sole purpose for visiting the website was to search for porn sites.

The strategy of users carrying out all of their online activities on a single portal was flawed from the start. To be fair, in the mid-nineties, it was difficult to predict just how far the internet could and would explode, but it wasn't impossible, either. The internet was rapidly uprooting traditional media. The capability to deliver content and advertising is more dynamic than traditional television or print ever could be. Yahoo! improperly thought that they would be the new medium's version of NBC, CBS, and ABC combined.

The media strategy was flawed from the start. While it's inarguable that there were components and similarities with traditional media, the underlying advantage Yahoo! had and missed recognizing was constant innovation afforded by its underlying technology.

Technology moves at breakneck speed and offers opportunities not available to traditional media, which by history and definition is slower and more bloated. An organization that leads with technology innovates at breakneck speed, while traditional media had to be dragged by their ears to innovate.

Building an organization that was simply an extension of traditional media rather than leveraging the technological advantages only the internet provided resulted in an inability to think outside the box.

We've seen this throughout the history of internet businesses. Amazon.com innovated bookselling and commerce, while companies like Borders Books and Sears were left flatfooted. Netflix innovated video-on-demand while Blockbuster was left in the dust, and Apple innovated mobile technology, while Dell, Gateway, and others were left with big boxes of useless wires and processors.

Google entered the market as a force for innovation. It's no wonder that it quickly surpassed Yahoo! as the single most relevant company in the history of the internet.

While Yahoo! was defining itself as a media company for the twenty-first century, Google had a simpler goal—to produce relevant search results. Google provided the front door, first for millions and then tens of millions and hundreds of millions and eventually billions of people on the internet. The company precisely pointed a user to where they wanted to be. It's like having a personal shopping guide at Target or Walmart. As someone who is *not* a professional shopper, I waste more time searching the Target aisles than actually driving the fifteen minutes each way to and from the store.

Ironically, by enacting the search strategy, Google quickly became "all things to all people." Yahoo!'s portal strategy was more analogous to being "*some* things to a lot of people."

In the mid- to late nineties, Yahoo!'s biggest competition came from long-ago businesses such as Excite, Infoseek, and Lycos. None is memorable today and each has gone the way of *The Old Farmer's Almanac*, *Encyclopedia Britannica*, Woolworths, parachute pants, and Members Only jackets.

Those were the days of the "Portal Wars." Each portal offered more-or-less similar services to Yahoo!, except our marketing strength allowed us to survive and build larger market-share while the others were eventually sold and swallowed into a larger company's black hole of nonexistence. Expect the same fate for Yahoo! post-Verizon merger.

While the portals were all following Yahoo!, no one was chasing Google. There was a parade of ships sailing into a storm of fiery hail, locusts, darkness, livestock disease, biting insects, and the rest of the plagues while Google was a parade of one with a clear eye toward a parted sea without the forty-year part of walking the desert. At least Yahoo! walked the desert for close to twenty years while the others died of starvation after only a few.

Believing that the way forward would be unicorns, butterflies, roses, and blue skies resulted in the stubborn conviction that they were a "media company" first and foremost. While company executives realized the Yahoo! service required a strong technological foundation to scale, it presented itself as a media company externally and acted within the walls of the old media rules internally.

Traditional media has had a long-standing and appropriate rule that editorial and advertising departments should always be separate. It's very similar to Republicans and Democrats today, without the appropriate part. Editorial integrity should never be compromised in exchange for paid advertising. However, in a new medium where it was unclear whether or not the search results were closer to editorial content or actual advertising, Yahoo!'s insistence on presenting itself as a media company pushed company leadership to incorrectly define search as "content." With search internally considered editorial content in the tradition of old media, it resulted in advertisers not being allowed to pay to manipulate how high their brand would appear in the results.

At the time, advertisers were given prime space at the top of the page, referred to as banner ads. Today it seems as antiquated as a mainframe computer the size of an entire floor of a building. The initial results were strong, but they were a novelty. As users realized these were ads, the click-through rate (the number of users who clicked on the ad to be sent to the advertiser's website) decreased substantially. While it was logical for a traditional media company to dedicate advertising space, it was short-sighted for an internet publisher to do the same.

First, the internet, by its entire nature, is the most efficient direct marketing platform ever created. While banner ads had a call-to-action, they were primarily branding exercises. A thirty-second television ad is a great place for Coca-Cola to brand its product with a compelling story accompanied by some godforsaken jingle like, "I'd Like to Buy the World a Coke…" Annoying, disturbing, tortuous, yes, but still quite effective. There was no way to tell a story in a banner ad and thankfully no way to hear an agonizing related song either. Advertisers were developing branding creative for what was essentially a small billboard, synonymous with a larger billboard on the side of a highway. People jumped from one page of a website to another as quickly as if they were driving a car by the sign. The difference was that you might have been fortunate enough to entice someone to stop and click-through to a more robust sponsored content experience that existed, but nevertheless those chances still remained minimal. Coca-Cola wasn't going to sell more sodas with a branding ad on the internet. Marketers attempted to get smarter, but how smart can you get if you've been doing the same thing for one hundred years? And now you are being prompted to try new ideas far removed from the business culture companies built over dozens of years of existence. Advertisers would tempt users to their websites with hokey games. After a certain point, AT&T's website offering trivia games of the history of AT&T just doesn't have that much viral appeal.

Take Procter & Gamble, an always forward-thinking, traditional organization that certainly tried to learn how best to leverage the internet for marketing purposes. It was one of the first companies to embrace the power of television, and the company knew it had to invest in the potential of internet marketing rather than stay on the sidelines. P&G's entire existence depended on developing and selling great products by creating thoughtful and memorable branding campaigns on television. It initially struggled, not surprisingly, to succeed in a banner advertising environment. Pampers, a Proctor & Gamble-owned product, obviously has a very narrow target audience comprised almost exclusively of expecting moms or moms with small children. P&G executives are very smart individuals, so my guess is they didn't have a lot of interest in speaking to me. I'm just not thinking about the differences of plushness and comfort between Luv's, Huggies, and Pampers. All I knew is that whenever the baby needed changing, my only goal was to change the diaper as quickly as possible to minimize my dry heaving. To this day, I don't have any recollection of the brand of diapers our children used. I just wanted the experience to end quickly and painlessly (for me).

Banner ads target individuals based on a larger comparable demographic set, similar to television and print except with an immediate call to action. This was Yahoo!'s bread and butter for years, and, for the most part, it has gone the way of the dinosaurs.

Online advertising today is highly invasive. Technology has vastly improved, and privacy policies favor the publisher. I'm calling it the "chip in your head" advertising model. Advanced machine learning allows algorithms to target users as if the publisher has installed a "chip in your head."

If I go to 7-Eleven and purchase a product, like Dasani water, I can expect to see an advertisement for Dasani when I log into Facebook or another online publishing company. It's as if my actions have been caught on camera in order to be immediately sold to the highest bidder. Now, I'm on the fence on that. On the one hand, I am terrified that marketers have invaded my privacy, but on the other hand, it's kind of cool to know that someone believes I have value. But the ethical angel on my shoulder usually wins out on this evil versus good debate.

The most effective online marketing will always be achieved through search results. There are two reasons for its effectiveness. First, it's the only advertising created that targets a potential buyer or customer 100 percent of the time when they are actually *asking* to be targeted, and second, because the universe of potential advertisers is opened to the world. When my Yahoo! team was selling banner ads, our paying customers were primarily either large traditional marketers or young internet start-ups that had raised boatloads of capital. Today, anyone who wants to sell a pair of vintage Chuck Taylor sneakers can list their product to market on the search phrase "Vintage Chuck Taylor Sneakers." The chance of finding an interested buyer is increased from .02 percent to 100 percent.

CHAPTER 9
Paid Search: The Largest Missed Opportunity in the History of Business—Take 2

While it's been heavily reported over the years that Yahoo! rejected the offer to purchase Google for one million dollars, what's been less established was Yahoo!'s role in increasing Google's service. Earlier, I explained that the original Yahoo! model was having humans review every website submitted and then carefully categorize each site into its proper topic, sub-topic, and sub-sub-topic, and so on. It was the Dewey Decimal System filled with rocket fuel.

Clearly, with the exponential explosion of the rise of websites, the company realized this was not a scalable model. However, building a search engine is a very expensive and time-consuming task. While Yahoo! was building a back-end search engine for its proprietary directory, it needed a web-based search solution (even if the search capability was being de-emphasized and management was considering moving the search bar to appear at the bottom of the page). A few years after turning down the acquisition of Google, Yahoo! had licensed their search technology; therefore, every search on Yahoo! was a search on *Google*. The unforeseen result was Yahoo! helped build its eventual largest competitor.

Additionally, tension existed between the product group and the business operations group, or what could fairly be described as the user side and the revenue side. It cannot be overstated that tension between the needs that users demand and the needs that advertisers demand is healthy. Neither side should ever have complete control of all decision-making. If that were the case and a sales department had responsibility over all final decisions on how ads should appear on the page, then the user experience would have consisted of highly annoying pop-up ads filled with heavy graphics, dramatically slowing the time for loading a webpage. Recall, back in those times most people using the internet from home were logging on through very slow modems or ISDN lines. High speed internet was just graduating from the Pong phase to the Atari phase (or, said another way, from infancy to crawling).

It made sense for Yahoo!'s product teams to push back on our sales teams. While we were responsible for driving revenue, we also were heavy users and understood that user experience should never be detrimentally compromised. At times internally, we were seen as the enemy to users, but in reality, the push-back from our department was minimal.

To the credit of the product team, they slowly evolved and allowed the sales team to be more active in product development. As new products were created, the Sales Program group that I led was the bridge between the product and the sales teams. The product leaders or producers, their official title, begrudgingly became less rigid. Jeff Mallett, president of Yahoo!, was starting to place new pressures on the product side to be more cognizant of the relevancy of our advertisers when developing new services. My Sales Programs team helped to forge a trust between the two primary operational sides of the company. This unwritten peace treaty essentially assumed the producers would be more respectful of the importance of generating ad revenue as long as the sales team understood the clear distinction of what was considered editorial versus advertising and would never compromise the user experience. Unfortunately, it was as if the Israelis and Palestinians had signed a peace agreement without ever answering the question of who has land and occupational control of Jerusalem. You just knew it wouldn't last.

While there was a truce on how banner ads could appear on the page, there was no agreement on how the company would proceed in the paid search environment.

In February 1998, serial entrepreneur Bill Gross launched a novel search engine called Goto.com. While the company has long been forgotten, a case can be made that it was one of the three most important innovations in the internet's early era, the others being the advent of the web browser and email. For our first few years of Yahoo!'s existence, we developed a financially successful program allowing advertisers to purchase banner ads on the top of the page of keyword searches. There was a Wild West aspect to this program. At first, a banner was sold on keywords for one thousand dollars per word, no matter how significant the search terms were in terms of usage. IBM brilliantly took advantage of this offer. In 1996, as the World Wide Web was taking off, they reached out to me to purchase an IBM banner on the search term "sex." This meant that each time a user typed "sex" in their search box, an IBM banner ad would appear at the top of the page. While I don't recall the exact number of times the word was searched per month, it was far and away *the* most-searched term at the time. The number of searches for sex and sex-related words and phrases is difficult to ascertain. Rather than releasing the top search words, search engines tend to release the top search trends. In 2016, according to Google, the top search trends were Powerball, Prince, and Hurricane Matthew. There is not a fiber in my body that believes for a second that the word "sex" and any sex-related terms were searched less than "Powerball," but that's a topic for a different (and definitely more exciting) book.

However, while Google and other search engines may hide and debate the top-searched words, there is no denying in the late nineties it was sex, and IBM smartly capitalized on that obvious market reality.

Other organizations purchased banner ads that appeared above a competitor's brand. Volvo bought the search word "Honda" or some other car company. We debated the issue internally and determined it was ethical and legal.

Often, it's the moments in our lives that we don't recognize that help shape the rest of our existence. It's like being in a bar and going up to a group of women, meeting one, falling in love, having kids together, and then divorcing five years later. What if you had chosen another bar that night? What if you had found a connection with her friend instead of her? It can become an exercise in insanity to attempt to determine a different fate to life.

This was what happened to Yahoo! in February 1998. It not only profoundly changed internet advertising but changed the fate of Yahoo! and Google forever. Bill Gross's Goto.com introduced the utterly wild concept to charge advertisers for their results to appear at the top of the search results, in text format, seemingly integrated into the results. The first true paid search program! As news spread throughout the day, it became the talk of the company. I viewed this model as the "Great Experiment" with revolutionary potential.

The Goto.com announcement was thrilling. It was the first definitive indication that search could be viewed through a monetary lens beyond selling antiquated banners on words such as "sex" and "music."

As the head of Sales Programs it was my responsibility for my group to determine new sources of advertising revenue for the Yahoo! network. It was clear that no other initiative could have generated as much new dollars for the company as selling sponsored search results to advertisers throughout the world.

There was one major obstacle. While the Sales Programs team was bouncing with joy at the prospect of creating an entirely new business model, the producers remained steadfast in their belief that search was always going to remain proprietary content and thus in no way could it breach the sanctity of the editorial versus advertising carbyne wall. (Carbyne is arguably the strongest material in the world.) But this wall wasn't just made of carbyne, it also was lit with molten lava and if you could penetrate the molten lava then you'd be electrocuted by a wall that was powered by a yoctowatt of electricity. (Yoctowatt: the approximate power of gravitational radiation, the strongest order of magnitude of power defined.) It's probably fair to acknowledge the product team wasn't going to consider heading down the path of sponsored paid search.

To say this was disappointing is an understatement. Whereas banner advertising was simply traditional media camouflaged on a computer screen, sponsored search was the differentiator that demonstrated the full capabilities of direct-targeted marketing businesses had always envisioned. It was *advertisingtopia*. It was as if the woman you've waited for your entire life suddenly appeared at your doorstep and threw herself at you. It was as if you could eat all of your favorite foods such as nachos, pizza, ice cream, Big Macs, Whoppers, and Baconators, and never gain a pound or feel full after a meal. It was as if you found the magic lamp with the genie but you had unlimited wishes, not just three. Sponsored paid search was truly the "Holy Grail" for marketing executives everywhere.

The Sales Programs team recognized this while the product team steadfastly believed selling search results would destroy the trust the company had built with its hundreds of millions of users worldwide. So, without the blessing of the producers, our team had to go back to putting a bowtie on a mayfly. Sure, we can dress it up but a mayfly has the briefest lifespan of any animal in the world. Without paid search, the end was just beginning.

Mike Moritz of Sequoia Capital has often been recognized as of one of the most successful venture capitalists in history. His knack for recognizing world-changing technology companies is unprecedented. His investments include some of the greatest brands in business today, including YouTube, LinkedIn, PayPal, Yahoo!, and Google.

In 1996, Moritz was the first venture capitalist to believe in Yahoo! and who infused the company with its first round of venture capital. In doing so, he became an original member of the Yahoo! board of directors.

A few years later Moritz made a similar investment in Google, betting that Sergey Brin and Larry Page had created the best search engine with the most relevant results. Google struggled early to identify revenue streams, and both Brin and Page abhorred banner advertising, Moritz strongly suggested that Google follow the lead of Goto to develop a paid search program.

The Google founders were intrigued and along with their top engineers and marketing executives rolled out a product called AdWords in late 2000.

AdWords was built on a similar thesis that Bill Gross had pioneered, selling pay-per-click text ads on search results. Typical of Google, however, they improved on the original paid search model Gross had developed two years earlier. Google's program was set up as an auction allowing individuals and companies to purchase ads directly through Google's website without any need of outbound sales. There were other advantages as well. Inherently the targeting was more precise; advertisers only paid when their ad was clicked through to their website or landing page. As presented exclusively as text, the cost of developing expensive banners ads was reduced to zero. Most important, the universe of advertisers was opened up to any small business or individual in the world wishing to market a product or service. In no time, paid customers grew from hundreds to millions.

The program was so successful that Google developed the next generation of AdWords, called Ad Sense, allowing them to sell targeted ads on third-party websites, literally millions across the internet.

With Google poised to dominate the paid search market, the Yahoo! product team stubbornly believed users would recognize the purity of Yahoo!

and abandon any search engine that sold result placements. They were wrong. Google successfully integrated advertising with search in a way that was accepted by the entire world.

The rule of separating content and advertising was well established. Writing the rules for search, as it turned out, was the most important decision in the early years of online monetization. Yahoo! determined that search was editorially sacred while Google determined that search was the sacred cow. (To be fair, while Google witnessed out-of-this-world success, users also always experienced the fastest loading and most-relevant search results.) The search decision altered the destiny of both companies and, by extension, the experiences and future online habits for billions of people across the world. It was bigger than the Red Sox selling Babe Ruth's contract to the Yankees in December 1919. The Yankees would go onto win twenty-seven World Series titles, while the Red Sox had to wait eighty-six years before they were champions again. It was bigger than IBM not recognizing the future of the desktop computer and selling their DOS operating system to Bill Gates and Paul Allen for fifty thousand dollars in the mid-seventies. It was bigger than Time Warner selling to AOL for 160 billion dollars in 2001, only to see the entire merger collapse a few years later. Just how big was it? Google AdWords and AdSense programs now generate close to *seventy-five billion dollars in annual revenue*. Yahoo!, in contrast, was so late that it wasn't until the third quarter of 2014 that earnings from search revenue outpaced display advertising when the company reported revenue of 452 million dollars. (In 2017, the company was sold to Verizon for more than 670 billion dollars *less* than Google's market value as of July, 2017.)

CHAPTER 10
The Storm Came

By late 2000, cracks were appearing in the market for internet stocks. At our previous summer sales conference, Anil had stood on stage and warned the company with his infamous, "A storm is coming" speech.

By 1999, Anil and I discussed our fear that our business was a house of cards ready to fall. The collapse of internet stocks surged in earnest the first half of 2001, and although we weren't surprised, there was nothing we could do to get out of the way of the approaching tidal wave. Had the product team agreed to paid search, it still would not have been enough to avoid seeing our revenue collapse. However, the fall wouldn't have been as drastic.

Imagine a tightrope walker performing on a wire thirty feet above the ground and then imagine three potential outcomes resulting from falling. The first outcome is falling into a net, still an unintended fall, but you'd be left uninjured. The second outcome is falling onto one of thin pads that were ubiquitous in high school gymnasiums across the country (I always found it odd that we weren't allowed to chew gum in school, but we were told to climb a rope thirty feet high in gym class with only a three-inch pad below to protect us!). The third outcome is falling thirty feet off the wire onto a bed of nails. Had Yahoo! pursued and embraced paid search, the fall would have been into the net. The company would have bounced back up and continued its routine unharmed. Early internet start-ups such as eToys, Pets.

com, Drugstore.com, govWorks, and others fell onto a bed of nails, never to be heard from again with their tombstone being nothing more than a small footnote in the annals of internet businesses. They were the mayfly. Up to that point, Yahoo! was a profitable business with hundreds of millions of loyal users. This saved us from the nails, but nevertheless our fall was still onto the thin pad. We broke some bones, endured lots of pain, but had enough assets that we didn't perish. We still had the chance to recover and eventually attempt to get back on the high-wire again.

Our fall was precipitous. In a matter of eighteen months our market value dropped from a high of one hundred and twenty billion dollars down to eight billion dollars. The internet companies that had spent millions of dollars on banner advertising on our service stopped paying their bills all at once. While we anticipated the crash, we were hamstrung to avoid it.

We knew that relying primarily on dot-com companies for our robust stock price was unsustainable. Wall Street analysts, the kind of people who "Main Street" believes are reprehensible scumbags, were quite successfully earning their reputation. The large investment banks and brokerage houses such as Merrill Lynch, Goldman Sachs, and Morgan Stanley would identify a company whose sole value was adding .com to their name. They had a highly unethical, though legal, strategy of securing millions of dollars in fees to underwrite the public offering while ignoring the fundamental fact that the business had no revenue and very few prospects for future success. The analysts on the other side of the investment house were recommending the stock to their banking clients to prop up the value. If ever a conflict existed, this was it; but this being Wall Street, conflicts are as common as arguments between siblings. Like all other nefarious Wall Street plans, this predictably went south and went south quickly. The result was dot-com companies filing bankruptcies at an alarming rate. Congress passed and President George W. Bush signed the Sarbanes-Oxley Act, which among several measures set new standards of conduct and regulations for the management and boards of publicly traded companies. The most regrettable result of the irrational exuberance and Wall Street greed was that millions of Americans saw life savings vanish.

For years, we had internally griped that stronger relationships had to be built with large traditional marketers. We needed to focus less attention on closing big deals with start-up internet companies and focus on initial smaller partnerships and more solid relationships with major companies such as Target, Wal-Mart, Pepsi, and other traditional advertising spenders. Our position as such a high-profile publicly traded company made it virtually impossible to look much beyond the upcoming quarter. In college, I was similar to most college students in that I'd cram the night before the exam. In our Yahoo! world, we knew the test was coming but even if we studied months or years in advance, we'd still fail.

Wall Street demanded that we exceed our quarterly revenue goals to the detriment of future earnings. Had we missed our quarterly goals, it wasn't outside the realm of possibility that the entire market would have crashed, causing small investors to lose their life savings and the value of the stock each employee owned to plummet. Imagine the pressure of kicking a last second field goal to win the Super Bowl, Buffalo Bills fans; imagine the pressure not to let the ball go through your legs in the tenth inning of the 1986 World Series, Boston Red Sox fans; and imagine the pressure of being the losing goalie in the semi-finals of the 1980 Olympic hockey tournament, Russian hockey fans! That's the immense pressure we felt for twenty-one straight quarters. The difference for us was, it wasn't a game. It was the life savings of millions of individuals and families.

By the end of the first quarter of 2001, it was impossible for the house of cards to hold itself up. The dot-com companies stopped paying their bills and there was nothing we could to hold the house intact. Tim Koogle, the original CEO, was forced to resign because the company reported poor earnings. Yahoo!'s earnings announcement reported that the company, for the first time in twenty-one quarters, would miss its revenue goals. In subsequent months, the Nasdaq would fall sharply by almost 80 percent from its high of 5,132 reached on March 10, 2000. Our fear that investors' life savings could be wiped out was accurate. "Irrational exuberance" caused the stock market crash. While dozens of tech companies were complicit with Wall Street, Yahoo!'s standing as arguably the highest profile earnings miss was among the most influential contributors to the market downfall.

CHAPTER 11
Terry Semel: Old World Failure

On April 17, 2001, Yahoo! announced that Terry Semel had been hired as the new CEO. Semel's very successful career culminated with his rise to become chairman and co-CEO of Warner Bros. Studios.

Within a few weeks of Semel's hire I resigned from Yahoo! I chose to leave for several good reasons, but in the end, I knew it was time. I'd accomplished as much as I could and had no interest in trying to prove myself to a new management team. Coinciding with the arrival of the new CEO, Anil had also resigned, and Jeff Mallett, another mentor, who was the inspiration and drive behind the first six years of Yahoo!'s existence, was going to be less influential in a Terry Semel-led regime.

When I walked out of the building the last day, I was proud of all my accomplishments but I did have two regrets. The first was that we hadn't developed stronger relationships with large, more traditional marketers, and second, that myself and the Sales Programs team I led had lost the internal debate to develop a paid search program.

The lifeblood of any organization is the continuation of the corporate culture that inspired its greatness from day one. Yahoo! had a first-rate culture. Employees arrived at work energized, long hours at the office went by too quickly, there was a shared belief in the purpose, goals were clear and achievable yet aggressive, co-workers became family, we learned from an endless collection of brilliant minds, and Yahoo! afforded us an ecosystem that allowed us to gain interest in topics miles outside our own experiences. What was most important and fulfilling each day was to have a role in changing the lives of the more than 7.5 billion people on Earth.

This was the Yahoo! culture. Even if we had different ideas on how to achieve our goals, each and every one of us came to work each day with the pride and excitement that we were making the world better. And we did. Yahoo! revolutionized the way the global population accesses information at its fingertips. That's a pretty good place to grow up!

By the end of my Yahoo! tenure, that culture had eroded. How could it not have? Revenue was dramatically decreasing. Google had demonstrated it was far more formidable than Lycos, Infoseek, Excite, or some other lost soul of a business we beat back with ease. Enthusiasm over earlier acquisitions, such as the Mark Cuban-founded Broadcast.com had been extremely disappointing, and most notably our stock price had cratered.

However, at the core of all the negativity was comfort knowing that with Jeff Mallet remaining and Jerry and David still intimately involved, not all hope was lost. My close friend Bob Greenberg had a history of world-changing success right out of dreams and fantasies. At Harvard in the mid-seventies Bob's arch nemesis for smartest kid in the class was Bill Gates. After graduation, Gates called Bob and asked him to join his early start-up in Albuquerque, New Mexico. Bob moved to officially become employee number four at Microsoft. One of the most fascinating and rarely told stories in the history of any technology business is that, very early on Microsoft had run out of money. Bill Gates asked Bob to front payroll for a few months. Bob agreed but he was repaid in cash instead of stock, stock that today would be worth over one billion dollars.

Among many other achievements, an early program Bob developed became the foundation for Microsoft Excel. In 1981, Steve Ballmer was hired and the company moved its operations to Washington state. Tension with Ballmer and a lack of interest in living in the Northwest resulted in Bob leaving Microsoft and relocating to Connecticut to take a senior role in Coleco, which happened to be the Greenberg family business. It's been reported that Bob played a significant role in recognizing the commercial potential of the most famous Coleco product, the Cabbage Patch Kids. One of the nation's great secrets is the same brilliant technologist, marketing whiz, and entrepreneur gifted to the world the twofer of Excel and Cabbage Patch Kids. Only Nolan Bushnell, who founded Atari *and* Chuck E. Cheese, compares as favorably.

Coleco was founded in 1932 as a company that produced shoe leather. In the mid-1970s the company successfully transitioned to being a toy manufacturer. After experiencing unprecedented growth in the 1980s by releasing the Cabbage Patch Kids, Trivial Pursuit, ColecoVision, and other memorable products, the company decided to enter into the home PC market. Companies such as Texas Instruments, Commodore, and then Apple were successfully validating Bill Gates vision of having a desktop computer in every home. Coleco released the Adam, their personal computer, in October 1983 to disastrous results. Serious drawbacks included such components as the loudest and slowest printer on the market (which is saying something considering how slow printers were in 1983) and cassette tape drives that easily unspooled, like an old VHS. By 1985, the Adam was discontinued, and by 1988, Coleco filed for Chapter 11 bankruptcy due in large part to losses from the Adam.

Bob always gives the same enlightening response as to why the Adam failed. A technology company had to be led by a technologist. At Coleco, Bob was the only engineer and he wasn't the company CEO. Bob knew full well that the Adam wasn't ready for release and made his case internally to no avail. In the end, Coleco released a product not ready for the market and that helped to crash the entire business.

All of the greatest technology CEOs have been engineers by training: Bill Gates, Mark Zuckerberg, Eric Schmidt, Larry Page, Michael Dell, Tim Koogle, and so many others. Terry Semel was clearly not.

A typical Silicon Valley work environment includes pool tables, foosball tables, free lunches, free soft drinks, open loft space or cubicles, electric skateboards, mattresses rolled up under the desks, and pets. The attire is so casual that an outsider would think there are daily "who-can-come-to-work-looking-the-most-homeless" contests, prizes to include a "Fat Elvis" poster and a lava lamp. Companies offer their employees free healthcare and 401K plans. In return for the flood of perks, employees work ninety-hour work weeks, sleep under their desks, live and breathe the corporate mission, and ultimately provide great results. That's a strong corporate culture!

As co-CEO and chairman of Warner Bros. Films, Terry Semel was by all metrics a very successful executive. Trained as an accountant, he was not particularly flashy, but possessed all the trappings of a successful Hollywood mogul. He amassed a fortune. He owned a large mansion, his own private plane, a personal driver, and had a reality show daughter.

Once Yahoo!'s board of directors had made the very unwise decision that they were going to be a media company, Terry Semel became a logical choice to become Yahoo!'s second CEO. History has demonstrated that going "all-in" as a media company was a flawed strategy. As described in previous chapters, media companies moved slower, had multiple layers of management, and were followers rather than leaders in innovation. The most negative element is that the corporate culture in a media company is traditionally as dry and dull as watching a documentary on the history of yarn.

Terry Semel was as exciting as background music on an elevator. When the CEO of a Silicon Valley company brings as much emotion as a toll booth operator, it sinks down through the rest of the organization very quickly.

Yahoo!'s board was under the flawed belief that it was time for an adult to take the reins of leadership. What they failed to see was that Tim Koogle was an adult. He not only brought gray hair from years of executive experience, but he also had a technology background so he could hold his own in conversations with the tech geniuses at the company. Jeff Mallett, while youthful and still in his early thirties, was an adult too. Jeff was an experienced Silicon Valley executive who inspired Yahoos to work long hours and buy into its world-changing mission. All Terry Semel brought was media experience and gray hair. Not only was he uninspiring but he had never built a corporate culture. Yahoo! was still just a six-year-old company while Warner Brothers had been in existence for decades. Let's not also forget that he had never been CEO of a publicly traded company. Koogle and Mallett had great success not only managing internal growth, but as Yahoo! executives they had been in charge of one of the highest profile public companies in the world.

Jeff Mallett was both engaging and approachable. He had the gift of making everyone in the organization feel as if they were a critical cog in the company's success. If Jerry was the external face of Yahoo!, then Jeff was the internal soul of the company. While he may not have always sided with us in an internal dispute (see "paid search"), we all understood we were important stakeholders to ensuring the mission was executed effectively.

Terry Semel, by contrast, was less approachable. He commuted from Los Angeles to Silicon Valley in a G5 Jet, he had a driver at all times, and had an office built for him that could have doubled as the Taj Mahal. He had no technical training, which made it impossible for him to converse with the engineering teams, and he rarely bothered to attend sales meetings with our largest advertisers.

Semel's expertise was in media and with the direction of the board, he initially set out to create a twenty-first-century media organization. One of his first decisions was to hire legendary television executive Lloyd Braun.

Braun was most notable for his time as chairman of the ABC Entertainment group, having given the green light to programs such as *Lost*, *Grey's Anatomy*, and *Desperate Housewives*. He also was involved in some of the most iconic television shows in history, including *Seinfeld* (the character Lloyd Braun is named for him) and *The Sopranos*.

Under Braun's tenure, Yahoo! moved its media operations south to Santa Monica. This set-up resulted in an internal clash of divisions of Bloods versus Crips proportions. It was Hollywood versus Silicon Valley. It was similar to Simon versus Garfunkle…David Lee Roth versus Van Halen…Brad Pitt versus Angelina Jolie or Democrats versus Republicans. (Actually, I take that back, no conflict is as awful as Democrats versus Republicans.)

Tensions over strategy turned to disgust. Kernels of truths evolved into fiery infernos that split Yahoo! into two cultures of single-minded independent organizations pursuing two unrelated paths. The war between the Southern California arrogance of an, "I'm right and you're an idiot" mentality versus the culture of Northern California built on an, "I'm brilliant and you're brilliant so let's go change the world together" mentality clashed so explosively that the entire corporate culture that Jerry Yang, Jeff Mallett, Tim Koogle, Anil Sing, and others so perfectly established deteriorated virtually overnight.

Stories spread that Lloyd Braun demanded his own private plane, converted a large conference room with a patio into his own private office, and reserved a parking space closest to the elevators. There was also the likely half-true tale of Braun stealing an umbrella from the company store. When the store clerk asked him for payment Braun infamously exclaimed, "Don't you know who I am?!"

While many of these incidents have been disputed by Yahoo! leadership and Braun himself over the years, the outrageous narratives portrayed him as a complete misfit to Silicon Valley headquarters.

Braun's responsibility beyond leading the media team was to develop network-type programming exclusive for Yahoo! Prior to this audacious strategy, Yahoo!'s video content consisted of showing either live events such as the Discovery space shuttle orbiting Earth on a flight mission or highlights of sports and news events from traditional newsfeeds. Braun, at the urging of Semel, launched a strategy to develop original programming that users watched only on Yahoo! on their computer screens.

Only in my warped mind can I recall otherwise forgettable television programs such as *The Brady Brides, After MASH, The Love Boat: The Next Wave,* and the personal favorite of five or six Americans, *Joanie Loves Chachi.* While clearly these programs were all cringe-worthy embarrassments, they achieved cult following because they were spin-offs of popular television shows. Yahoo!'s slate of original programs, including *Kevin Sites in the Hot Zone, Blog for Hope,* and the immortal *Richard Bangs Adventures* series, aren't even memorable enough to be parodied.

This strategy failed less because the programming wasn't memorable (ESPN built a cult following when it launched mostly log-cutting competitions and badminton) but was due more to inadequate systems necessary to bring the concept through each aspect of the production process. The strategy was shortsighted mostly because users were unaccustomed to viewing original programming on their computer screen. It would be years before YouTube shifted consumer behavior viewing habits to other screens, such as mobile devices and laptops, but even so, Netflix, the leader in original web-based programming, didn't take off until the introduction and mass production of smart television sets in the early 2010s.

While the media strategy was flagging, the corporate culture was deteriorating. The executives in the Yahoo!'s Santa Monica media headquarters were more interested in growing existing media properties, such as sports, news, and entertainment, while at the Santa Clara headquarters, the engineers were driven by the promise of developing new products and services. Santa Monica wanted to expand existing worlds while Santa Clara wanted to create new worlds. These tensions caused small cracks in the company to grow into massive canyons.

Is Yahoo! a media company or a technology company? The company could never agree on this central question. There wasn't a single CEO in the history of the organization who effectively directed the company toward a single strategy. Failing technology initiatives, such as search, and failing media initiatives, such as original programming, only resulted in more confusion and unanswerable questions on the proper path for the organization.

The revolving door of executives during the Semel years and the corporate culture war caused by different mindsets and abilities was taking the sand out of the weights of the company. There was an overall unhappiness from constant media scrutiny and a lack of a single strategy. By the end of 2002, Google was skyrocketing as the clear leader globally in search. Two bets that Yahoo! made early on proved to be disastrous. First, the vast exploding growth of the internet made it irrational for the company to de-emphasize search and try to keep users exclusively within the pages of the Yahoo! network. Second, there was the lack of realization that search was the most scalable way to increase revenue.

While the concept of original programming never took off at Yahoo!, products such as Yahoo! Sports, News, and Finance were still leaders in their respective categories; thus, there was an argument to continue and enhance the media strategy. Yahoo! still couldn't recognize that paid search had become significantly more relevant than display advertising banners on content pages. The explanation was that corporations who had spent years trying to learn how to best market their products online had finally caught up. Large ad agencies had either dedicated teams or they acquired advertising firms that exclusively focused on internet advertising, which by then had evolved into a cottage industry. They now had the knowledge to confidently convince their clients where and how much to spend their ad dollars on the internet. Unfortunately for Yahoo!, and fortunately for Google, display advertising had fallen out of favor.

CHAPTER 12
All Hands on Deck for Search: The Largest Missed Opportunity in the History of Business—Take 3

As previously examined, display advertising simply could not target customers of corporations as specifically as search engines could. It's similar to knowing the neighborhood a house is in, versus knowing the exact address of the house. The neighborhood can certainly indicate estimated demographics of the residents, but knowing exactly where a resident resides and knowing their regular routines and habits is clearly more valuable to the advertiser. The result for Yahoo! was it became more difficult to monetize extremely popular categories within their network.

In defense of Terry Semel, he recognized that Yahoo! needed to develop a search strategy quickly. As CEOs often do, when Semel was hired, he recruited some of his loyalists from his previous career at Warner Bros. One such hire was Jeff Wiener, who post-Yahoo! was hired as the CEO of LinkedIn, a company he eventually sold to Microsoft for $26.2 billion. Weiner was responsible for the entire search platform and therefore creating a strategy for search to compete with Google.

There was internal tension on how to proceed to develop a search strategy. Company leaders and, eventually, Semel himself saw the urgency and therefore leaned toward buying a company to help them compete, while the programmers favored building an internal solution.

A common Silicon Valley phrase is "Not Invented Here" or "NIH." NIH is used by programmers to presuppose in-house development is superior to what's available outside the organization. "Not Invented Here" arises most often during internal debates on whether to build a product internally or to purchase a company with the product already built and distributed. Programmers are like Olympic athletes; they are supremely confident in their talents. The best have the ability to go into the boss's office, light a cigar, put their feet up on the desk, and tell him or her—not ask—what their salary and company perks will be. Most important, they always believe in their ability to build a better product than what already exists somewhere in the market. In 2003, Yahoo!'s brand was still pristine, even with the internet bubble crashing two years before. As a result, the company still had the cachet to recruit top technical talent. The New York Yankees team has barely won one championship in the past fifteen years, but the team recruits the best free agents, because even in defeat, the brand is so strong that it has the resources to offer the highest salary and the cachet of wearing the Yankees pinstripes is unlike any in baseball. Yahoo! was like the Yankees, their best days behind them, but the Yahoo! brand was still strong enough that they could recruit the best talent and offer the best compensation.

With Semel as the final arbiter, the decision was made to acquire Goto. com, which in October 2001 had been renamed as Overture. Two years later, in October 2003, Yahoo! acquired Overture Services Inc. for $1.63 billion.

Remember: Yahoo!'s origin was not in search but in categorizing websites submitted to the company. The team of web surfers who lived in the dark worked day and night reviewing countless websites to determine the proper placement in the proper category.

Clearly, the number of sites developed in the late 1990s snowballed to a level that was impossible for the web surfers to manage. It was apparent that Yahoo! needed a search partner and until 2003, they partnered with companies such as Alta Vista, Overture, and Google to provide web-based search results.

The decision to acquire Overture signified Yahoo! was going to dive head first into building a search engine. Recall there are two types of search listings, one that is a paid search and the other that is a free or "editorial" listing. Yahoo! had acquired the search company Inktomi in early 2003.

Inktomi was a web-crawler-based search engine founded at UC Berkeley in 1996. The company previously partnered with Yahoo! to provide the software to power Yahoo! Search.

Additionally, Overture had acquired AlltheWeb (a Norwegian company with a search thesis similar to Inktomi that was even more highly regarded than Google in its early years) and Alta Vista, a search pioneer, which like Inktomi before and Google later, was a high-powered search engine that had once powered Yahoo! Search.

Yahoo!'s challenge was to combine the paid search aspects of Overture with the web-crawling technologies of Inktomi, Alta Vista, and AlltheWeb to form Yahoo! Search.

In April 2003, Yahoo! Search was officially launched. A notable outcome of the launch was the termination of their agreement with Google to provide search results. This was the red line in the sand confirming that the two companies were now bitter rivals. Yahoo!'s previous, highly consequential decision to de-emphasize search had allowed the two organizations to work collaboratively.

History demonstrates that Yahoo!'s decision not just to minimize search, but to partner with Google to provide search results, was a driving force in Google's unprecedented growth. Every Yahoo! search was powered by Google; therefore, Google *was paid* each time a user searched a keyword on the Yahoo! network.

The student had clearly surpassed the teacher. It's as if the student ended up driving a Bugatti Veyron 16.4 Super Sport, the world's fastest production car with a top speed of 268 mph, while the teacher was still driving a moped. The difference was as stark as the student riding aboard Triple Crown winner American Pharoah while the teacher was on a pony ride at the Louisiana Sweet Potato Festival. The gap became as large as the student's rise to president of the United States while the teacher was losing a campaign for assistant secretary/treasurer of their local homeowners' association. This difference was about $670 billion in market capitalization at the time Yahoo! was sold to Verizon.

In 2002, four years after Yahoo! turned down acquiring Google for one million dollars and blew one of the greatest business opportunities ever, another chance arose to buy the search leader. (I will always believe missing out on taking free water and selling it in a bottle for an infinite mark-up was, and will always be, THE biggest missed opportunity.)

In the summer of 2002, Yahoo!'s revenue was almost four times that of Google. The difficulty was that Yahoo!'s stock still hadn't recovered from the dot-com bust and Terry Semel, still an internet neophyte, had *no idea* how to value a technology company. He offered three billion dollars and the Google founders promptly turned down his offer. Internally, executives tried unsuccessfully to convince Semel to raise his offer. Semel resisted. Now, there are anecdotal reports that Google wasn't interested in selling for any price. (I'm sure if there had been a number that blew their board of directors away, an agreement would have been possible.) I always believe that everyone has their price. It's not known if five, seven, or ten billon dollars would have convinced Google, but somewhere there was (likely) a number that they would have accepted.

I'm convinced that if Tim Koogle and Jeff Mallett were still leading Yahoo!, they would have offered an amount that Google couldn't have refused.

After all, they had purchased Broadcast.com for $5.8 billion and GeoCities for $3.57 billion when both companies were generating minimal revenue. As an aside, although both acquisitions were considered failures, the thesis proved accurate, just premature as Broadcast.com was anticipated to evolve into what YouTube eventually became and GeoCities was the earliest social network. One good thing came from these deals. Mark Cuban, co-founder of Broadcast.com, made billions and was released to world.

What Semel wasn't capable of understanding was that tech valuations were by their inherent nature irrational. As someone originally trained as an accountant in Hollywood, he had a very linear mindset, and could never fully integrate into the Silicon Valley environment.

Missing out on acquiring Google the first time was understandable. Even if Yahoo! had emphasized search, there was no way to predict that Google would have become the enormous success and world-changing company it eventually became. It's also important to keep in mind that Sergey Brin and Larry Page, who have become two of the most successful executives in corporate history, originally had no interest in commercializing their search engine and were eager to return to their studies. Finally, had Yahoo! acquired Google, minimal effort would have been exerted to leverage the technology to innovate search, keeping in mind that Yahoo!'s strategy was to keep users within the Yahoo! network. Essentially, had the deal actually happened, there never would have been a Google; most observers agree that Google has helped improve the world in countless ways.

However, the same theory doesn't apply to the final attempt to acquire Google in 2002. By this time, it was becoming apparent even to technology neophytes like Terry Semel that search had scalable value. Google was on the precipice of becoming a verb and every programmer in Silicon Valley was convinced that the search technology Larry Page and Sergey Brin had built was far superior to other market-developed products.

Once the prospect of merging with Google was lost forever, Yahoo!, having implemented its search strategy with the acquisition of Overture and Inktomi, dug itself into a deeper hole answering the question, "Is Yahoo! a media company or a technology company?" Terry Semel's original goal was to present Yahoo! as a new media company for the twenty-first century, which meant it lives on a computer screen. With the Lloyd Braun original programming experiment going nowhere, Semel had no choice but to figure out a search solution. This clearly bent the company back toward technology.

What Yahoo! discovered once it acquired Overture was they had bought a very inferior product to Google. It was if they'd bought a large house in a beautiful neighborhood, but the home's design was flawed. The floor plan was poorly drawn and the walls were filled with asbestos. Google, in contrast, had built a beautiful mansion on pristine land, equipped with all the state-of-the-art technology, concrete floors, hurricane-proof windows and doors, and a roof that no wolf would ever blow down. Most exciting of all, they planted two trees in their backyard that grew unlimited amounts of money. They named the trees AdWords and AdSense.

Unlike Google, Overture was built to make money first and provide relevant search results second. Google was created to generate relevant search results and as a result, the AdWords system could generate incomprehensible amounts of money by combining paid search with relevant searches to increase the click-through rates. Overture had an antiquated strategy of displaying textual ads on search exclusively on the bid price of the per-click an advertiser was willing to pay. While Google went for quantity *and* quality, Overture focused only on quantity.

Additionally, Google had developed the systems for customers to purchase ads through a fully automated system, while Overture's search engine had yet to create that functionality.

Ironically, the Overture service was designed exactly in the manner that Yahoo! producers had rejected when the idea of paid search came up back in 1998 because it clearly compromised the user experience. Google, in contrast, was able to offer paid search combined with relevant search results and actually improved the user experience.

Due to Yahoo! having an inferior product Google was able to skyrocket ahead of Yahoo! Search in overall value. By 2005, Google had more than double the revenue of Yahoo! Search and was still growing exponentially, quarter after quarter, while Yahoo! was still attempting unsuccessfully to figure out the proper strategy.

This didn't go unnoticed by Terry Semel. In 2005, two years after the launch of Yahoo! Search, he assembled a team led by Jeff Weiner to develop a product to compete with Google, whose lead in search was growing faster than Jack's beanstalk. The project was called "Panama."

The year 2005 was the most critical year for Yahoo! It can be argued that, although they had fallen far behind Google, there was a chance to close the gap. Additionally, Yahoo! and Google were, for the most part, the "only game in town" to compete for ad dollars. The industry had not quite recovered from the dot-com collapse four years earlier but it was strengthening. Other mid-2000s sensations most notably included Facebook, which launched in 2004 and was just being introduced to consumers outside of colleges. YouTube hadn't gained significant traction and Twitter hadn't been created yet. By 2006, Facebook, YouTube, and Twitter had all become household brands with hundreds of millions of users, and they were starting to compete for ad dollars. By the end of the 2000s and into the early 2010s, companies such as Snapchat and Instagram had further crowded the market competing for ad dollars. The most critical was probably the launch of the iPhone in 2007, which created an entirely new platform for advertisers to spend their marketing budgets.

It was as if the U.S. highway system had been built in 2005 when only two cars were on the road. By 2006, a flood of new cars caused traffic to slow and by the 2010s traffic was at a standstill.

But in 2005 Yahoo! still had a chance, and although late in the game, Semel knew it was the last shot. Unfortunately, Panama, a grand and necessary project, was a disappointment. Several factors contributed to its failure. First, it was a highly complex project. An entire financial management system had to be built. Functionality had to be developed for advertisers to bid in an auction process, and a new user interface to place those bids had to be created. A reporting system allowing advertisers to track and optimize results also had to be developed from scratch. Whereas Google had already developed all of those processes and services, Yahoo!/Overture was light years behind.

The best way to overcome technical obstacles is to challenge programmers with complex projects. The best programmers possess both passion and the mindset that no technology problem is so great that it cannot be solved. As labor intensive as the Panama project was to execute, ultimately it was built and released. The greater challenge for Panama was internal bureaucracy and a culture clash between Yahoo! and Overture.

The importance of having a durable corporate culture is critical to the mission, and if corporate culture is feeble it can bring an entire company to its knees. The most extreme example of culture clash was AOL's purchase of Time Warner. The tens of thousands of employees trained in traditional media who were comfortable with layers of bureaucracy were never going to join in singing "Kumbaya" with an online upstart whose employees wore jeans and t-shirts, were given free lunches, and drove one hundred thousand dollar cars. AOL had twice the stock value and half the revenue. By 2005, Time Warner's incoming CEO had declared the merger with AOL as the "biggest mistake in corporate history." Almost ten years after the merger, Time Warner spun off the remaining assets of AOL.

This is not a book detailing the AOL-Time Warner merger, and certainly there is enough written on that subject to fill a library, but it does demonstrate just how fragile merging two very distinct cultures can be.

Yahoo! had a history of failing acquisitions. During my tenure, there were several big flops. Often, it's due to the acquired company not wishing to integrate themselves into the culture of the acquiring company. I've witnessed organizations that were producing one-tenth the amount of our revenue join with an air of superiority, which always baffled me.

An example of a poor culture fit was Broadcast.com. After Yahoo! acquired the company in 1999 for $5.7 billion, I was sent to their Dallas headquarters to meet their sales staff. Yahoo! was the high-flying, high-profile internet darling. Our annual revenues were closing in on one billion dollars. Broadcast, in contrast, had revenues of less than thirty million dollars. As I've noted, the purpose for acquiring a company with so little revenue (and for such a large amount) had more to do with predicting the future of what the organization *could* become. Yahoo! made a logical bet that Broadcast.com could be what YouTube eventually became. Unfortunately, with bandwidth still limited and high speed internet and mobile platforms still years away, this hypothesis was premature. However, another argument for failure was the culture clash. Broadcast had a sales force that was highly experienced in selling technology. On my first visit, I encountered experienced sales executives who had absolutely no interest selling Yahoo! display banner advertising. Why? Technology sales traditionally have much longer sales cycles, are more consultative, have an audience that is comprised of chief information officers, and making a sale requires a deep understanding of how the technology works. Media sales, in contrast, operate on quick sales cycles, are less consultative, have an audience that is comprised of marketing directors with minimal technology experience, and making a sale requires nothing deep about understanding how display banner ads work.

It didn't take long to see that there was no way these two sales organizations would be a good fit. Technology sales executives loathe media sales executives; they consider them to be inferior. To media sales executives, a technology sales process is completely foreign, from the products to the audience to the sales cycle. It was clear that some (but to be fair, not all), would have no interest in ever selling Yahoo!'s products and services. As my career progressed and I developed more experience in the technology sales environment, I became familiar with and was better able to understand the mindset of the tech sales executive. But even so, that didn't matter to me. We had acquired them, they needed to integrate into our culture, our revenue was more than ten times their revenue, and our market capitalization was almost twenty times larger. Being on the sales side, I had less of an opinion of the product and engineering teams, but my guess is the rest of Yahoo! shared the same frustrations that I discovered. In fact, I'm sure that was the case because by 2002, much of Broadcast.com's services were shut down and most of its employees were terminated.

By contrast, in 1997, Yahoo!, in need of an email solution, acquired Four11, which had developed a leading email service called RocketMail. RocketMail was a huge success. Not only did it quickly vault Yahoo! as the leading email provider but its top executives and staff integrated beautifully with and enhanced the Yahoo! culture. Their top executive Geoff Ralston eventually even became chief product officer for all of Yahoo!

One merger that was called off due to a culture clash was between Yahoo! and eBay. Following the AOL-Time Warner merger, there was pressure for Yahoo! to conduct a large, game-changing deal. In March of 2000, Yahoo! had agreed to purchase all assets of eBay for almost forty billion dollars. Both boards had voted in favor of this decision, and a press release was prepared. The deal was ready to be announced to the world. I clearly remember waking up early at home in Saratoga, California to turn on CNBC in anticipation of watching the announcement live. To my shock, there was no announcement. I discovered years later that the deal had cratered the previous day because Yahoo!'s founder, David Filo, had decided to vote his shares *against* the proposed deal, citing unease in merging the two cultures. I applaud David for recognizing the two organizations, while a good fit on paper, would not have fit well in practice. The Broadcast.com failure was unpleasant but not a highly tragic transaction. Had a Yahoo!/Ebay merger failed, it had the potential to ruin both organizations.

Corporate marriages and traditional marriages share some similar aspects but also have some differences as well. Neither party should ever move forward simply because it plays well on paper. Both types of "marriage" require chemistry; having a heartfelt, soulful, and spiritual connection; a commitment to support each other to grow and mature together; and an appreciation of their differences while leveraging the similarities.

The main difference is, in a traditional marriage, both sides are inspired to move *toward* each other to meet in the middle. In a corporate marriage, the primary responsibility of integrating cultures and beliefs lies with the acquired organization. It's less of a coming together in the middle than it is of coming closer to one entity, that of the acquiring entity.

A dominant reason mergers fail is because neither side takes the time to perform the proper due diligence to determine if their corporate cultures can co-exist harmoniously. Merger discussions are mostly negotiated in secret by the top executives. When announced, thousands of employees are left flat-footed with unanswered questions and an unpredictable future. Mid-level managers are usually given no notice to carefully prepare their staff or the staff of the acquired business. The psychological three-dimensional effect of a merger is de-emphasized in favor of an "it works on paper" two-dimensional view.

Yahoo! and Overture were filled with enough cultural brick walls to sink an ocean liner. There was the culture of Southern California/Overture versus Northern California/Yahoo!; the acquired company was years ahead of the acquiring organization in search, thus creating a sense of superiority; it was a technology organization versus a (new) media organization; entrepreneurial spirit versus bloated bureaucracy; the one-time undisputed digital organization pioneer rapidly losing ground to a once-smaller competitor (Google); and finally Wall Street pressure creating tremendous expectations that the acquired organization could not comprehend.

Each of these obstacles could have caused the entire merger to flop, but the combination of all of these negative factors all but ensured a titanic failure.

By 2005, there were over ten thousand employees at Yahoo! Abraham Lincoln may have grown government more than any other president in history, but he had the excuse of the Civil War. Terry Semel grew Yahoo! in a way that would make Republicans all cry mercy, but he didn't even have the excuse of a nation splitting in two (which is ironic, because Semel was one of the largest Republican donors in America).

The organization's headcount had ballooned to more than 3.5 times the size of the company when I left in 2001. It was apparent that Yahoo! was still a growth company. Developing necessary processes without creating a choking bureaucracy is arguably the greatest challenge that a CEO will face in a growth business. And one of the great failures of the Semel regime, which was another bullet causing the ultimate death of the company, was his inability to create the systems needed while not compromising the entrepreneurial spirit that internet companies need as fuel. He had taken a flat organization where many leaders were empowered to make decisions to a bloated reporting system where decision making was all top down.

Jeff Mallet, in the early years, had organized the company into three distinct departments (excluding legal and accounting): Product, Engineering, and Sales/Marketing. The Product Group was the hub of the organization. Each product had a distinct team comprised of a general manager, producer, assistant producer, and a business development lead. In subsequent years there were several reorganizations, as would be expected of a growing company. At the end of my tenure, products were categorized into three distinct divisions: Media, Communications, and Commerce. The Media team produced content heavy products, such as sports, news, finance, and entertainment. The Communications team produced products such as Yahoo! Chat and Instant Messaging, and the Commerce team produced products such as Yahoo! Shopping and Auctions.

In a mature organization, brand groups are each tasked with the responsibility of "owning" their own product P&L (the income statement that reports the profits and losses of a company). The result is each product group manages its entire business, including revenues and spending, as if it were an independent operation. For example, Pampers, Tide, and Pringles are all treated as separate businesses reporting into Proctor & Gamble, the parent corporation. This model works well when the different brands have no co-alignment.

The unique nature of Yahoo!'s network of services made it difficult for the sales force to be confined to selling programs specific to each service. The company had organized the sales teams regionally into East Coast, Central, and West Coast sales. For most of my tenure, Anil decided that each individual sales person would continue the development of programs across properties so there would be a team of generalists rather than individuals who gathered a deep understanding of a specific category. This made sense because user traffic was still generated primarily by news, sports, and finance. More niche areas for advertisers to secure a large audience to which they could market, such as autos, were slow to build a strong enough user base.

There was one exception to this rule, the Yahoo! sin team. While it was kept quiet, it was no secret that adult content, i.e. porn, is the largest generator of search traffic, not just on Yahoo! but all search engines. Considering Yahoo! had by far the largest audience, our traffic patterns indicated just how perverse the minds of individuals are across the globe when they are home alone. Anyone operating a porn site preys off of this sickness and makes boatloads of money off users' perceived privacy. Note to online porn watchers: Privacy is a fallacy! Somewhere in the cloud someone is watching you!!

The sin team was a small team selected to manage relationships exclusively with adult content companies that advertised their services within adult keyword searches. And yes, revenue goals for this team were among the highest in the company.

But I digress...back to boring organizational challenges. This system allowed Anil and the sales force to have complete revenue responsibility over all products. A positive was that the product teams had no dis-incentive to coordinate and collaborate across services. The downside to this model was that the product teams, with no revenue responsibility, had little incentive to develop products with the advertiser in mind. That would change and arguably a change was needed. As more accountability of the product teams was demanded under Semel, corporate tension was unleashed. Producers had a sole, narrow-minded vision to monetize their specific products, while the sales organization was agnostic to any one specific service with the sole goal to provide programs that were aligned with their client's strategies.

Terry Semel was a systems-driven CEO. With a history in traditional business, he rightfully demanded that each product team own their own P&L statements and other additional metrics, such as the amount of traffic to each website within the Yahoo! network. The different teams each grew from a few staff to literally hundreds of individuals. Heads of each group led product, business development, engineering, and design teams beneath them. The teams became so bloated that subsequent CEOs were pressured by activist investors (more on this later) to layoff 50 percent or more of the staff. While no one would or could argue that the product teams needed more accountability, how to do so remained a corporate challenge for Yahoo!'s leaders.

In a traditional brand-driven business, these challenges are less profound. *Sports Illustrated*, *People*, and *Fortune* each have a different culture but are all within the Time Warner family. They share or compete very little. Each is run as a separate business and a staff member refers to him- or herself as an employee of *People* or *Fortune*, not of Time Warner.

For large portals, the challenges are more profound. Sales executives never referred to themselves as working for Yahoo! Travel or Finance. They just work for Yahoo! The media buyers wanted to purchase ads across the Yahoo! network and most often not one service. Because of the advanced analytics and targeting capabilities, today this is less of an issue as advertisers use automated systems to maximize cost and efficiency across search engines and the need for explaining nuanced display advertising programs is archaic.

However, for years, and prior to Project Panama, Yahoo! struggled to determine the right processes. While Semel was correct in his plan to make each group more accountable, in an internet media company it's much more nuanced than the MBA 101 playbook.

Not only had Google automated the processes, but they constantly reinforced their strategy to be a technology company with their focus solely on the goal to provide the best search listing for its users and advertisers. Goals are much more likely to be achieved with less clutter and less competition. Yahoo! was still all over the map as to whether it was a media company or a technology company. Although internally the company was heavily focused on the Project Panama rollout, leading media properties still required significant attention.

Which all leads back to Project Panama. To his credit, Semel recognized how important this endeavor was and made it a top priority for the company. Hundreds of millions of dollars of resources were allocated to see it succeed, but it needed more than just money. The entire organization needed to pull in the same direction instead of competing for dollars and mindshare internally. Semel's heavily bloated organization and his accountability processes caused considerable delays to Project Panama. While I previously examined the cultural challenges of merging the expertise of Overture and Yahoo!, there was an equally vexing obstacle that the combined search organizations were not getting the support they needed from other internal departments. For example, if the search team wanted to advertise on the home page, it had to submit a request to the homepage team and then it was considered for approval by a committee. Often the request was denied. In the minds of the guardians of the Yahoo! home page, which at the time was the most-visited single page on the World Wide Web, any internal promotions would result in revenue they couldn't recoup from paying advertisers. Their goals were fixed and there wasn't any incentive to collaborate. It was as if the Secret Service denied the Secretary of Defense, Secretary of State, and Secretary of the Treasure access to the president.

Over time, Semel calmed the cultural war by merging the Overture teams and the Yahoo! search teams. Jeff Weiner was given responsibility to oversee the entire project and, in the end, he developed a product that was well reviewed by the market. Unfortunately, it was too late. The entire project, which was conceived in mid-2005, wasn't launched until early 2007. The new and enhanced Yahoo! Search Marketing Service was delivered at least two business quarters too late. By the time it was released, Google had opened up a sizable lead in the search market. Google had evolved from a search engine to a verb, literally listed in the dictionary. With a unified corporate mission, they weren't standing still. They enhanced their product significantly. Most notably they improved their targeting capabilities for advertisers. With increased traffic, Google had a full realization of not just how best to target users, but which keywords performed the strongest. Advertisers on Google, which numbered in the millions, were seeing far better results.

By the end of 2007, Google had won the search wars while Yahoo! was back to the drawing board trying to determine where to turn next. Google's stock was soaring while Yahoo!'s was floundering. By 2008, Google's stock was valued at 180 billion dollars, while Yahoo! was worth thirty-five billion dollars.

CHAPTER 13
Facebook: The Largest Missed Opportunity in the History of Business—Take 4

Unfortunately for Terry Semel, falling flat on the promise to deliver original programming, losing out on acquiring Google, and delivering an underwhelming performance on the Panama Project were not his only mistakes. Missing out on a golden opportunity to acquire Facebook has to rank near the top as well. Eschewing sarcasm, if not acquiring Facebook was *only one* of his biggest blunders, then it's astonishing that the company survived as long as it did.

As of July, 2017, Google is valued at 680 billion dollars and Facebook is valued at 475 billion dollars. I'll go out on a limb and state with confidence that Terry Semel is the only CEO ever to overplay his hand enough to turn away two companies that would ultimately be valued at greater than one trillion dollars combined. That's probably not the legacy most of us strive for.

Facebook launched as a dorm-room website in February 2004. Four months later, founder Mark Zuckerberg dropped out of Harvard, moved his company to Palo Alto, California, and received his first outside investment when Peter Thiel, co-founder of PayPal, acquired 10.2 percent of the company for five hundred thousand dollars.

This book is not about how Facebook started and grew and that is now, along with Google, enjoying world domination. That story's been covered more than the national anthem has been sung in schools across America. What's fascinating is to look at a side-by-side comparison between Yahoo! and Facebook, which, in its thirteen-year history, has yet to make a significant game-altering error. During the same time, Yahoo!'s history has been filled with enough mistakes to fill California's lakes during their epic ten-year drought.

Facebook was a high-flyer from the moment Mark Zuckerberg hit the "go" button in his dorm room. Within a few months after launch, Facebook moved their headquarters to Silicon Valley. Companies came knocking on their door to learn about the social network that was connecting college students all over America. Facebook was not a pioneer but that wasn't relevant. MySpace, Friendster, and others had already launched social network platforms and each had millions of users. Facebook's secret sauce was its brilliance initially targeting college students. Technology was their religion, the campus was their church, and Facebook was their revivalist spiritual leader. Their model was better than any that Jehovah's Witnesses could ever pray for. Instead of wearing out shoe soles and getting doors slammed in their faces, Facebook got millions of students to convert with a click of a mouse.

Over the course of its first two years, Facebook had no limit on the number of suitors that wanted to purchase the company. As a venture capital-backed business, the goal of investors is to show a return. Make no mistakes, these are very smart individuals and they make deals to sell companies in their portfolio only when timing and opportunity align. For Facebook, that moment was mid-2006.

Prior to that moment, there were other opportunities. Friendster, Viacom, MySpace, NewsCorp (the MySpace parent company), and even Google made inquiries. In some instances, there were real offers that, for a number of reasons, did not lead to an executed deal. Viacom, sensing Facebook's synergies with MTV Network, which they owned, was told "no thank you" during its initial discussions. However, intent on acquiring the social network, Viacom returned a few months later with an offer of 1.5 billion dollars. This was an astounding amount to offer a company less than two years old showing minimal revenue. The board of Facebook clearly took the offer seriously, but ultimately rejected it because seven hundred million dollars of the $1.5 billion was to be paid out at a later date.

At a certain point, the Facebook board had instructed Mark Zuckerberg to sell Facebook if he received an upfront one-billion-dollar offer. Although Viacom's offer was more, the fact that part of the deal was based on a future earn-out was not enough to convince Facebook to sell.

In early 2006, Yahoo! entered into discussions to acquire Facebook for 1.1 billion dollars. Facebook had many of the same assets that Yahoo! cherished. They had a growing and connected audience. Their user-generated content was addictive. The public wasn't just visiting Facebook; they were *staying*, while Yahoo! was more like a 7-Eleven, where you can stop in for a coffee, a Big Gulp, or a Slurpee and leave. (I consciously excluded hot dogs on a roller in this example because I have never, and will never, stop in to 7-Eleven to buy a hot dog on a roller). Yahoo!, of course, would never openly acknowledge that any user ever left its website. Google, on the other hand, is more like a moving sidewalk at an airport where an individual is effortlessly led to the proper terminal and exits the sidewalk. In 2006, Google openly acknowledged this. And beyond just getting you to your gate, Google provides maps and directions to your next travel destination. Facebook's different. They were the house that no one ever wanted to leave. In fact, in time they've grown from a house to a building to an arena to a city to a state to a country and eventually to a planet—where no one wants to leave.

As an extension of traditional media, Yahoo! was slow to recognize the power of user-generated content and user-shared content. In the world Yahoo! created, content was presented to its users from reputable publishers.

The only difference from traditional print or newspaper was that content on Yahoo! was refreshed more often and delivered via a computer screen. Facebook turned this paradigm on its shoulders. It demonstrated beyond a doubt that internet users were much more interested in reading thoughts and comments from their friends (and strangers) and sharing them across their network.

Yahoo! essentially had two theses upon which it was based that Google and Facebook proved, without any doubt, were completely misguided. The first was that no one would ever want or need to leave the Yahoo! network. Score that invalidation to Google. The second was that users want a one-way relationship with content providers. Score that invalidation to Facebook. After blowing the chance to acquire Google, a higher power dropped on Terry Semel's lap a present so fortunate that it had to have been gifted directly from the hearts of the Greek gods and goddesses.

Timing and opportunity aligned. Before June 2006, Facebook rejected acquisition efforts because the investors valued their company higher than the offers presented. Post-June 2006, Facebook, still operating a social network for college students, opened their doors and welcomed the entire population of Earth. I imagine it would be similar to Willie Wonka allowing the entire world to enter his candy factory instead of just the six lucky Golden Ticket winners. This allowed Facebook's value to skyrocket overnight, and by 2007, an investment by Microsoft had valued Facebook at fifteen billion dollars. By this period, Facebook's interest in selling had diminished to near zero. The board and Zuckerberg already imagined the tens and ultimately hundreds of billions the company would be worth.

However, for that one moment a deal was possible. And guess what? Shocking, but Yahoo! and Terry Semel blew it. Knowing that a deal could be done at one billion dollars upfront, the two sides had a tentative agreement. The potential acquisition was applauded by analysts with Mark May of

Needham & Company describing the arrangement as a win-win for both companies. May's highly sound theory was that as Yahoo! released new products the Facebook audience would be a perfect platform to introduce them, since college students tend to be early adopters. In return, the sheer size of the Yahoo! audience would instantly grow the Facebook user base by hundreds of millions almost overnight. Obviously, as it turned out, Facebook required no assistance to grow their audience.

Unfortunately, just as the merger was entering its final negotiations, Yahoo! reported weaker than expected second quarter earnings. Semel, ever the accountant and MBA playbook CEO, used this news to lower his offer to eight hundred million dollars. Facebook walked. Zuckerberg, still in his mid-twenties, would have made tens of millions in the transaction, but as it turned out no one was happier. Yahoo! overplaying its hand allowed Zuckerberg to go back to his board having rejected the offer because it was below the one-billion-dollar threshold. As it turned out, that was the last time any company ever had the opportunity to purchase Facebook.

Considering no one could have predicted the astronomical rise of Google, there is justification for Yahoo! rejecting an offer to acquire the current search behemoth for one million dollars. While less defensible than the first buying opportunity, the second go around, when Google was a more established organization, the two companies still found themselves several billion dollars apart on valuation. The negotiation to purchase Facebook, in contrast, put the companies at only three hundred million dollars apart on an agreement to acquire what arguably can be considered the most significant global network in history. I remain convinced that had Jeff Mallett been negotiating this transaction, the deal to acquire Facebook would have been cemented.

But just how bad did it turn out to be? Crazy stupid bad would be an understatement. As of this book's publication, Facebook had a market value of almost 430 billion dollars more than the value at which Yahoo! was sold to

Verizon. In addition, Facebook's total 2016 revenue was $27.6 billion dollars while Yahoo! earned $5.1 billion dollars during that same period. And lastly, Facebook long ago surpassed Yahoo! as the leader in display advertising.

It's important to provide additional context or at least a contrasting view. Yahoo! has a history of poorly integrating acquisitions. As you recall, both Broadcast.com and GeoCities, one-time internet high-flyers, were purchased by Yahoo! for more than nine billion dollars combined. They were monumental failures, although there was a multitude of defining factors explaining why poor planning of integrating employees and services into the parent company proved to be significant considerations.

CHAPTER 14
Flickr: The Largest Missed Opportunity of an Internal Division in the History of Business—Take 1

B ut of all the companies Yahoo! didn't miss out on and actually did acquire, none was as big a screw-up as Flickr. Not because of its price—the transaction was for a measly twenty-two million dollars—but rather because unknown to Terry Semel and his team (I know, it's getting really repetitive), Flickr was the first great social network and, if it had been seeded and watered properly, it had the potential to be as powerful as Facebook and Instagram combined. Flickr was founded in 2004 by the now legendary internet entrepreneurs and one-time husband and wife team of Stuart Butterfield and Catarina Fake as an online photo and video community... but their company was acquired by Yahoo! in 2005. Sound familiar?

The all-too-familiar cry of bloated bureaucracy, risk aversion, and blinded vision contributed to a world-changing missed opportunity. For most companies, that spells death. For Terry Semel's Yahoo!, it just spelled Tuesday.

Stuart Butterfield, when asked about his Flickr experience under the mothership of Yahoo!, articulated his clear frustration on how hard it was to get resources, "because of Yahoo!'s internal 'screwed-up-ness.' "

Securing investment and resources from Yahoo! management was driven by the financial success of the product. Highly profitable properties such as Yahoo! News and Sports weren't begging for their next crumb. Flickr, unfortunately, was. The strategy was so misguided, but it was right out of Terry Semel's "Dummies" book on "How to build the proper culture where risk is penalized and bureaucracy is rewarded." It's acceptable to invest in profitable businesses if your name is General Electric or Procter & Gamble or any other company formed around the time of the Jurassic Age. A growth company, especially an internet company, is, by its definition, *growing*, which means new products; even some considered "with risk" require significant resources. It's allowable, even actually expected to invest in business units as either loss leaders or potential growth engines. Facebook spent nineteen billion dollars on WhatsApp even when the company was generating minimal revenue. (As an aside, WhatsApp was co-founded by Jan Kourn and Brian Acton, both proud members of the Yahoo! alumni.) Unfortunately for Butterfield and Fake, Yahoo!'s bureaucracy was completely blind to identifying its potential within.

It's important to answer why Semel purchased Flickr and then chose not to provide the resources the organization required to scale. Yahoo! did so because it sensed images had potential to be monetized. While that's certainly accurate, it completely missed the entire of mission of Flickr, which was about building and serving a community. It's the difference between sending a rocket to Mercury to find money versus sending a rocket to Mars to find life. What Facebook demonstrated was that, if you identify and grow life, you'll ultimately discover money—and lots of it. If you're just searching for money and not life, then what's the reason to have money in the first place?

To demonstrate its depth of misunderstanding, the Yahoo! team ordered the Flickr users to abandon their log-ins for a universal Yahoo! log-in. That is completely disrespectful to the Flickr community. Individuals go where they feel welcome, where their voices are heard, and their friends are nearby. That Yahoo! failed to recognize the three primary adages that create the unbreakable foundation of community was another terribly regrettable mistake in the era of Terry Semel. As of April 2017, according to Amazon.com-owned Alexa Internet, Inc., a global website ranking service, Flickr ranked as the 352nd most-trafficked website while Instagram ranks as the eighteenth most popular website globally and Facebook ranks at number three. (Neither statistic includes users on mobile platforms. If included, Instagram and Facebook's share would be much greater.)

The thesis, of course, is it's more likely than not that had Yahoo! acquired Facebook in 2006, the company would have languished, halted its innovation, and would never have ended up changing the world by connecting over one billion individuals across the globe.

The Terry Semel reign finally ended June 2007, a little more than six years after his hire. His tenure was memorable for all the wrong reasons. He began with great fanfare, but in the end, he was cursed by two foundational obstacles that he wasn't able to overcome. First, he was a non-tech-savvy individual leading an internet company, which is one of those square peg/ round hole fits; second, he never was able to answer the question of whether Yahoo! was a media or a technology company. In the end, Semel's weaknesses resulted in a bloated, bureaucratic organization filled with preposterous missed opportunities the scale of which has never been equaled in business before. By mid-2007, Yahoo!, the once high-flying internet and stock market darling, had, under the watch of Terry Semel, forever lost search and social on the internet.

CHAPTER 15
Jerry Yang: Founder CEO

With the Semel years officially closed, the question of how Yahoo! was going to pick up the pieces was still unknown. Jerry Yang, co-founder and the reason I was hired, was announced as the third CEO of the company in June 2007.

While I never met Terry Semel, I knew Jerry well. He was a very important figure in my life and without him and Anil, I would never have achieved the unexpected heights my life reached. When Yahoo! went public and the stock value soared, Jerry became the first billionaire and then multi-billionaire I had ever met. It wouldn't surprise me if *he* was the first multi-billionaire he had ever met prior to Yahoo! either.

In the mid- to late nineties, he had become a true celebrity. He was featured in every magazine, often on the cover, and every newspaper in America and beyond. Overnight he became an internet legend in the United States. As an Asian-American, he was revered throughout China and Japan as well. With all of the attention he was generating, Jerry remained very approachable and "mostly" humble. Whenever he visited our New York office, he would make sure to take our team to lunch or dinner. On my first day at Yahoo! in the old Mountain View, California, office, I fondly recall going to lunch with Jerry where he taught me how to use chopsticks. When I moved and worked in the headquarters office, this billionaire would sometimes rummage through my desk searching for the upgrade tickets I had earned as a frequent flyer for his next trip. There wasn't a single Yahoo! employee to whom he didn't show devotion, and his love for the staff was reciprocated. He walked the halls of the company and knew the names of every Yahoo! employee worldwide.

Jerry was much more than a corporate cheerleader. He was intimately involved in every decision in the company and he was genuinely conscious of his responsibility to build a great company for his employees, customers, and shareholders. Jerry and Mallett—seemingly to the outside world—were attached-at-the-hip partners. They shared a vision for the company and executed their strategy. (Unfortunately, as has been demonstrated throughout the years, the strategy of being a portal was flawed.).

After Tim Koogle was forced out in 2001, Jerry was instrumental in the hiring of Terry Semel. From 1996 to 2000, Yahoo! was a run as a successful triumvirate of Tim Koogle, Jeff Mallett, and Jerry Yang. Each was complementary to the other, which worked well while the company was growing at lighting speed. Jeff was the operations guru, running all things day-to-day at Yahoo!., Koogle was the grey-haired veteran who had the confidence of Wall Street, and Jerry was the outside face of the company to the media as the company's chief evangelist. His official title was "Chief Yahoo!" David Filo, Yahoo! co-founder, was quiet and reserved. He led all decisions related to the Yahoo! infrastructure but was loathe to get involved in corporate politics. He had zero interest in being the face of the organization. He wielded extensive power internally, but only when necessary.

The relationships started to show cracks around early 2000. AOL had famously acquired Time Warner in January of 2000. Mallett, Jerry, and Koogle, on the heels of this world-changing development, had to decide if they wanted to follow AOL's lead and purchase a large media company such as Disney or stay the course to be the leading business on the internet. (How crazy does that sound today? A six-year-old internet start-up thinking about acquiring Disney!) After consideration, it was determined, and which can be argued rightfully, that Yahoo!'s best bet was not to merge with a larger organization such as Disney, Viacom, or NewsCorp. However, the pressure to do a big deal was ever-present. The AOL-Time Warner merger had led Wall Streeters, in their never infinite wisdom, to call for Yahoo! to make a big splash. It was decided by the triumvirate that they would approach eBay, a company with which they had on-and-off discussions for years about a potential acquisition.

On paper, this was a dream merger. Both companies were internet high-flyers. Their strengths complemented one another beautifully. Online shopping was slowly gaining acceptance. Amazon.com had evolved from selling books to selling the world. Most important, consumers were gaining comfort in using their credit card online. eBay was the undisputed king of auctions. Yahoo! had flirted with an auction site to compete, but the results were underwhelming. eBay would have provided experience, a shopping audience, a commerce, and a payment platform. In return, the Yahoo! network could have provided limitless marketing reach to hundreds of millions of users worldwide.

In hindsight, this merger would have paid dividends during the dot-com bust in 2001. Advertising is very much like stepping on a vehicle's gas pedal: When times are bullish, more pressure is applied; pressure is taken off the gas during more bearish periods. Marketers spend in good times and retreat in bad. In 2001, ad spending inevitably decreased because times were bad. eBay's business model proved to be more recession proof. Shoppers look for bargains in downturns and nowhere were there better bargains to be had than on eBay. As Yahoo!'s revenues were dropping, eBay's revenues were growing more than 70 percent in 2001. This was when the rest of the dot-com businesses were crumbling as violently as an asteroid falling on a sand castle.

Even with obvious synergies, the merger failed to be executed at the last moment. In most businesses, it's clear who is in charge. At Apple, Steve Jobs was CEO and the undisputed leader. Meg Whitman was the CEO and the undisputed leader of eBay. Carly Fiorino was the CEO and the undisputed leader for Hewlett Packard. Within Yahoo!, Tim Koogle did not have the same respect as Jeff Mallett. As CEO, Koogle wasn't aloof like Terry Semel, but his involvement in the day-to-day operations of Yahoo! was less apparent.

Within the walls of Yahoo!, Tim Koogle was the CEO but Jeff Mallett always seemed like the undisputed leader. I can't recall one time in my five years with the company that I had to seek approval from Tim Koogle. More interestingly, I can't recall a single decision that Anil, Yahoo!'s EVP and worldwide head of sales and marketing, had to seek Tim Koogle out for approval. In the hundreds of meetings I had with high-level executives from major companies, I don't remember more than two or three that he attended. For context, these meetings were CEOs of organizations such as Costco, Procter & Gamble, Prudential Insurance, and so many others. (I do have to give Koogle credit for skipping the meeting with the Enron CEO.)

Toward the end of my time with the company, when it was obvious our revenues were cratering, Koogle would seek me out for updates on the sales force and their progress. It was odd. While, I was an executive, I was still three levels beneath the CEO. Jeff, Anil, and even Jerry, my bosses, were highly tuned into the deepest minutia and all capable of providing answers. It didn't take my consultation with one of our triple PhD MIT engineers to figure out that Jeff and Anil were simply ignoring him.

Internally, almost everyone at Yahoo! believed Jeff would be the next CEO. I'm sure he would have had unanimous support among all staff. Jeff was loved and respected internally. He was hired as the senior vice president of operations, promoted to chief operating officer, then president, and was appointed to the board of directors. Clearly, Koogle, to his credit, had set up what seemed to be a seamless CEO transition. Unfortunately, as it happens in so many companies when the sky starts to fall, the small cracks that are unnoticed in the good times grow to become large sinkholes that will swallow anyone in sight. These aren't the average, annoying potholes left as remnants of a long and cold winter. These are holes more of the order of the ground opening up and dropping anyone within its radius deep into the Earth's liquid core, with temperatures of five thousand degrees Fahrenheit.

The relationship between Koogle and Mallett deteriorated after the eBay debacle. Had the merger gone through, it actually would have represented a very significant threat to Jeff's rise to CEO. Meg Whitman, eBay's CEO, was widely considered one of the best tech leaders in Silicon Valley. After a distinguished career at Disney, Proctor & Gamble, Bain Capital, and many other recognizable organizations, she was hired by eBay in March 1998 as the first "outside" CEO. Her tenure there was remarkable. In ten years, she grew the organization to over eight billion dollars in annual revenue, and throughout this period became one of the most powerful executives in the world. She had tremendous respect internally, but more importantly, Wall Street was enamored with her.

It was clear to any observer that Tim Koogle was not going to be the long-term CEO of the merged organizations. Mallett was right to conclude that had Yahoo! acquired eBay, it was entirely possible that the combined boards would have ultimately elevated Meg Whitman to CEO. Less than forty-eight hours prior to the deal being announced to the world, Mallett approached eBay's Brian Swette with ideas on how to share power in the combined organization. Meg Whitman would report directly to Jeff rather than directly to Koogle as had been decided. The problem was that Swette was a Whitman loyalist and he brought his discussion with Jeff to her attention. At the same time, David Filo, a Mallett loyalist, announced he would vote his shares against the deal. Both of these incidents conspired to kill the deal the weekend prior to it being disclosed.

While a case can be made that a corporate culture war could have ensured the sinking of the combined entity, the unfortunate timing made Mallett the scapegoat. Within twelve months, the dot-com bubble hadn't just burst; it was as if the Red Sea—which had remained parted for five years—suddenly shut. Most internet companies were washed away. Only a small few businesses such as Amazon, eBay, Google, and Yahoo! (albeit with serious injuries) made it safely to the other side.

With Koogle's (forced) resignation in March 2001, most observers were surprised that Jeff Mallett was not elevated to CEO. As it turned out, Jerry and the board surprised Jeff by informing him they felt he wasn't ready to be CEO. Mallett was still only in his mid-thirties but had, by all observers, been credited with the tremendous growth of Yahoo! and for good reason. Unfortunately for Jeff, the fact that he overplayed his hand in the eBay negotiations, along with overseeing the slide downward during the crash, forced Jerry to rethink who had the right profile to be the second CEO of Yahoo! Terry Semel had become somewhat of a mentor to Jerry. The two originally met at Herb Allen's media conference in Sun Valley, Idaho. The Allen and Co. Sun Valley Media Conference is an invitation-only event held each July. It's notable for its attendees, comprised of the biggest names in media, technology, politics, and sports in the world. It's the type of room where, if mirrors were placed on the walls, no one would ever leave. All the soldiers of the Roman Empire aren't enough to provide security for the greatest global collection of the world's richest, most famous, most successful, most fascinating, and most narcissistic men and woman on planet Earth.

The two men grew to admire each other. In Jerry, Terry Semel saw a young technology visionary who had created a company that had overnight changed the world. In Semel, Yang saw a highly successful executive who had overseen tremendous growth and had the kind of gray hair that commanded respect. Most notably, they could bond over their upbringing. Although Jerry was from Taiwan and Terry was from Brooklyn, neither were children born of silver spoons means. In fact, they were more likely born with zinc spoons, the least valuable metal on the planet. Both rose to incredible heights through vision and tenacity to be true change agents for consumers across the world.

Jerry convinced the board to hire Semel in May 2001, passing over Jeff Mallett. All reports indicate that Semel worked hard toward forging a relationship with Mallett and that Jeff was grateful for those efforts. However, any CEO has the full right to surround themselves with their own teams, which Semel did. This turned out to be the stellar executives of Jeff Weiner and Dan Rosensweig. Additionally, CFO Sue Decker—already a respected Yahoo! leader when Semel was hired—quickly asserted herself as a chief confidante to the new CEO.

On April 4, 2002, almost one year after Terry Semel's hire, Jeff Mallett resigned. As one of the most influential internet leaders ever, he was responsible for growing Yahoo! to over one billion dollars in revenue with more than one hundred and twenty billion dollars in value and offices in twenty-seven countries. Not a bad legacy of running a company for five years.

This brings us all back to June 2007 when the board of directors elevated Jerry Yang as the third CEO of Yahoo! Steve Jobs reappeared at Apple and forged perhaps the greatest turnaround in corporate history. Mark Zuckerberg was thriving as the CEO of Facebook. Jeff Bezos was creating the greatest retail company in the world. Marc Benioff, CEO of Salesforce.com, and Larry Ellison, CEO of Oracle, built two of the largest IT enterprise corporations. Bill Gates had arguably held the title of the most successful technology CEO of all time. The point is that corporate boards had to be blind (which at times the Yahoo! board has been accused of being) not to clearly see a growing trend in founder CEOs.

All of the names listed, as well as Google founders Larry Page and Sergey Brin, who along with Eric Schmidt acted as co-CEOs, were titans of success.

Before birth, when they were in line to fill their DNA, God loaded them all up with unlimited amount of drive, vision, and risk-taking. Steve Jobs visualized what consumers wanted before they ever knew it. Mark Zuckerberg's risks include acquiring WhatsApp—without revenue—for nineteen billion dollars. Larry Ellison had sales skills that rival Lincoln's convincing Congress to adopt the Thirteenth Amendment. The Yahoo! board evidently believed that Jerry possessed the same characteristics as other successful founders.

They were right and they were wrong. There should be no question about Jerry Yang's drive. As a leader, along with Jeff Mallett and Tim Koogle, he built a herculean brand known worldwide with over a billion monthly visitors. Jerry was instrumental in every critical decision. He traveled a million miles annually to open offices throughout Europe, Asia, South America, and Australia. He'd attend a sales meeting in New York on a Monday, and on Tuesday, fly to a meeting with key Yahoo! investor Softbank in Japan. When he was working in the home office, he was the last to leave. Sadly, it was a misguided drive that ultimately led to his most public failures (more on that topic later).

Along with his relentless ambition, he possessed an unwavering passion for Yahoo! No one has ever evangelized the Yahoo! brand more convincingly. Other founders move on and leave their creations for others to grow and run. Often, they achieve success in other ventures. Elon Musk, co-founder of PayPal, is now recognized primarily for being the founder of Tesla and SpaceX—and the alter-ego for Marvel's Iron Man character. Peter Thiel, another PayPal co-founder, launched a successful Sand Hill Road venture capital fund and was the co-founder of Palantir, a private data and security company that's been rumored through unconfirmed reports to have had a significant role in identifying the whereabouts of Osama bin Laden. He also was the earliest investor in Facebook. With all that Thiel has achieved, he is primarily recognized for underwriting Hulk Hogan's successful lawsuit against *Gawker*—which bankrupted the yellow journalism organization—and for his apparently close relationship with Donald Trump, most notably when he became the first and highest profile Silicon Valley executives to publicly endorse the future president. It's not often that Elon Musk and Peter Thiel are still heard discussing PayPal.

Jerry, on the other hand, was all in at Yahoo! He had other interests, such as golf, but he was exclusively connected with Yahoo!, unlike none other, including his co-founder David Filo, who was more comfortable coding in his cubicle with the other engineers.

However, whereas Jerry competed with every other founder in drive and passion, it's certainly arguable that he lacked the vision of a Mark Zuckerberg, Elon Musk, Steve Jobs, or Bill Gates. Gates envisioned a PC on every desktop. Jobs envisioned making computing as small and portable as imaginable. Zuckerberg envisioned connecting the world. Musk, of course, is on a whole different level. He envisions the Milky Way in the year 3017 with flying cars, mind-embedded chips that turn our thoughts into reality, time travel across rips in the universe's black holes, and a future where Vulcans and Romulans live in perfect harmony.

Jerry's vision wasn't even initially to create Yahoo! As a hobby, he and Filo had categorized websites on the burgeoning World Wide Web, generated users, gave their website the catchy name of Jerry and David's Guide to the World Wide Web. It wasn't until he and Filo recognized the volume of users was an important indicator of commercial viability. The power of the internet isn't just about creating a useful way to categorize websites, such as how many squares the average turtle has on its shell or finding a site dedicated to the original cast of *Gilligan's Island*, but rather the power is to *connect* people, information, and ideas across the world. Without a vision for the full potential for this groundbreaking, life-altering, world-changing (both positively and negatively) new tool called the internet, it was impossible to compete with the likes of the Jobses, Zuckerbergs, Bezoses, Pages, Brins, and others.

Lastly, Jerry lacked the risk-taking attributes of his fraternity of "founders." Ironically, this wasn't the case during the Mallett years. Yahoo! took several risks, some successful and some not, which is the nature of risk taking. Not all work out but those that do can bump the fortunes of a company from average to exceptional overnight. In the 1990s, Yahoo!, led by Jeff and Jerry, took several risks that seem routine today but were groundbreaking then. For example, Yahoo! was the first internet company to buy a thirty-second television ad during the Super Bowl. Jerry gave the okay for the iconic "Do You Yahoo!?" ads. He opened offices and dedicated in-language Yahoo!-branded websites in countries all over the world. Most notably, he drove the organization to launch Yahoo! products, such as Email, Finance, and Sports, that hundreds of millions of individuals across the world integrated into their daily habits.

It was only when Terry Semel arrived that the company became less risk adverse. There was no doubt that there were lingering fears over the dot-com crash, but I'm convinced that had Mallett been CEO with Jerry as his partner, Yahoo! would have acquired Facebook. Mallett took risks; Semel, for the most part, did not. Jerry, it can be argued, followed the risk makeup of the company's operating head. It is also possible that because Jerry was the most instrumental board member both in Semel's hiring and Mallett being passed over, that he was less likely to push back hard on Semel's decisions.

CHAPTER 16
The Evil Empire—Take 1

Ironically, Jerry's passion for Yahoo! was his greatest contribution to the company, yet it was also his biggest curse. After nearly nine months as CEO, the floor started to collapse beneath Jerry. Again, this wasn't a small pothole; it was a massive sinkhole that stared straight into Earth's inner core. In early May 2008, Steve Ballmer, CEO of Microsoft, called Jerry to advise him that Microsoft was going to be making an unsolicited hostile takeover of Yahoo! The companies had flirted in merger discussions as far back as 2006, but little came of it. The lackluster reception of Project Panama combined with Microsoft's futile attempts to develop its own search product resulted in Google furthering and eventually cementing its dominance in the search engine space.

The technology landscape of 1995–1999 is almost unrecognizable today. Google—spelled Googol—is a mathematical term defined as the number one with one hundred zeros after it (or ten to the power of one hundred). It was also the title of the not-so-famous 1913 children's book by Vincent Cartwright Vickers, creatively titled the *Google Book,* which detailed the rollicking adventures of the "Google" and his fanciful creature pals who live in the merry world of Googleland. As an aside, I had recalled Google being a peanut butter spread, but upon further research, the brand was called Koogle, introduced by Kraft Foods in the early seventies. Thus, for the truest of conspiracy theorists, perhaps the first Yahoo! CEO, Tim Koogle, did have more to do with the creation of Google than we initially suspected. I'm confident that the only family in America that consumed Koogle was the Ring family—which probably led to a market collapse a tad less publicized than Koogle's Yahoo! At least the peanut butter was kosher.

Mark Zuckerberg was a young teenager during this time. While I presume he was already programming in C++, I'm sure at that age, as a physics whiz, he was more likely trying to figure out how to drop an egg from five stories up without the shell breaking. Steve Jobs was just returning to Apple from his exile to the magical land of cowboys, space rangers, potato heads, and piggy banks. A "tweet" was a bird with an annoying speech impediment that drove some cat straight to the loony tunes, and Instagram was something you would get a co-worker for their bachelor party.

In the late nineties, the undisputed corporate champion was Microsoft, with zero losses and a bankroll that would make Augustus Caesar blush. (For those who don't know, Caesar ruled Egypt from 63 BC–14 AD. His wealth in 2015 was estimated at 4.6 trillion dollars because he owned all of Egypt.) Not only was Microsoft the highest-valued company in the world by the late nineties with a 1999 market value of 586 billion dollars, but it spread fear to "googols" of businesses worldwide.

Of all the cautionary tales of Microsoft's dominance, none is more illustrative than their absolute smash-down of Netscape. They did to Netscape what Sherman did to Atlanta or Reagan did to that medical school in Granada. They destroyed it with one massive weapon: Internet Explorer. Netscape not only pioneered the web browser, but their innovation was the first consumer window into the World Wide Web. If someone wanted to argue that Netscape's creation was the most important innovation in the history of technology, it would be hard to disagree. At their height, the Netscape Navigator web browser was installed on over 90 percent of the PCs in the domestic market. Microsoft, while late to recognize the power of the internet, quickly shifted its corporate strategy. With Microsoft Windows in total control of the screens of personal computers worldwide, they were able to release their browser, Internet Explorer (IE). Nearly overnight, they owned the browser market, while Netscape was left to reattach its toes, fingers, nose, ears, and so on. (Of course, we don't cry too much for Netscape. It was able to sucker AOL into buying its legacy technology for over four billion dollars.)

For years, we were warned, "Don't be *Netscaped*!" This rallying cry became a verb inside Yahoo!

Jerry disliked Microsoft. He made no secret of it internally. Other companies were competitors, but Microsoft was the "Evil Empire." There were varying opinions on how to proceed with them, as partner or foe, but we all were careful and often reminded ourselves not to "awaken the eight-hundred-pound gorilla." It's ironic that in the early years they mostly ignored trying to compete with us. Yahoo! had demonstrated without a doubt that advertisers would spend big money on search engines, yet it wasn't until 2009 that Microsoft went all in with the launch of "Bing"—its search engine designed to compete both with Google and Yahoo! Before Bing, Microsoft had only dabbled in search. In 1998, they launched MSN Search with Inktomi providing search results. In 2006, Windows Live search debuted. It was a vast improvement on MSN Search, but it still generated limited interest.

Steve Ballmer, who had known Bill Gates at Harvard, began his Microsoft career in 1980 as the thirtieth employee of the company. He held several management positions, becoming president in 1998 and eventually succeeding Gates as CEO in 2000. Microsoft employees described Ballmer as arrogant, rigid, hard-charging, and lacking vision. However, others described him as a good partner to Bill Gates, a brilliant manager, and a genius marketer. Microsoft's early culture was vision without discipline and order. The company was built and run by techies before Ballmer's arrival.

175

His tenure as EVP of sales and support and his time as president were by all accounts extremely prosperous. Microsoft grew to be the highest valued company in the world, and in the process made Ballmer a billionaire many times over. The primary difference between a chief operating officer and a chief executive officer is that a CEO requires a high business IQ along with a strong vision, while a COO's task is to execute the vision of the CEO. And, although Ballmer was president, his role was consistent of that of a COO. Therefore, Ballmer's lack of vision as COO masked the challenges he would endure when he was promoted to CEO.

In the late nineties, Microsoft was the undisputed leader because Bill Gates had a very rare combination of being both a visionary and a brilliant business executive. By 2008, Microsoft looked like a tired company more in line with IBM and Hewlett-Packard than with Google or Facebook. Speed-to-market was the hallmark of Microsoft's culture. The difference between Steve Jobs and Bill Gates was that Jobs would never release a product until, in his estimation, it was perfect, while Gates released products filled with problems with the intention of fixing the bugs in future product releases.

Under Ballmer's leadership, Microsoft's speed-to-market became the speed of a turtle. Microsoft lagged in embracing the internet. While it was easy to catch up against Netscape, more nimble organizations such as Google were crushing them in search and introducing products such as Google Docs to bite into the once-insurmountable lead of Microsoft Office. Microsoft never expected to lose its tight grip on its ownership of the computer home page. Only a few years earlier this would have been unthinkable, but no one ever thought the Roman Empire could crumble either.

Additionally, Microsoft had the scars from its late nineties and early 2000s battle with the government over anti-trust violations. At one point in 2000, a federal judge even ruled to break up Microsoft, only to be reversed on appeal a year later.

As an aside, during the Microsoft antitrust period, Ballmer had suggested to Yahoo! that we take over the operations for MSN, its travel site Expedia, and its leading email service Hotmail.

Lacking innovation and wounds from years of government hostility had taken its toll, and for almost a decade during Ballmer's tenure, Microsoft's stock was flat.

Steve Ballmer and Terry Semel had other similarities in that neither was considered an enormous visionary. Ballmer was more technically savvy. At least he could read his own emails, but Semel at least recognized Google's threat a few years earlier. Even so, what both companies had in common is they were getting their clocks cleaned. While Semel made a half-hearted offer to acquire Google when Google was acquirable, Ballmer waited until Google was the clear leader before he developed a strategy—a failed strategy—to compete. Like Semel with Project Panama, Ballmer sensed Google's dominance, albeit late, and tried to counter years later with an all-hands-on-deck corporate initiative to build the Bing search engine. But before the Bing release, he shocked the world with his hostile takeover attempt of Yahoo!

The stars did not align much during Jerry Yang's short tenure as Yahoo! CEO as he was entirely absorbed opposing the Microsoft coup. A consequence of his resistance to the hostile acquisition was the door to the cage opened for activist investors to spread plagues across Yahoo! for the rest of its existence.

By 2008, Google cemented itself as the dominant force in search. Microsoft, arriving even later than Yahoo! to the search wars, had one significant advantage over all other tech companies, except Google: they had the stock and the capital to buy their way into contention. Bill Gates had a philosophy that Microsoft should always have enough money in the bank so that the company could afford to go at least one year without selling a single product. Therefore, the means to acquire Yahoo! was apparent. The biggest obstacle for Ballmer wasn't the cost, it was Yahoo!'s public insistence that they were worth more than the forty-four billion-dollar offer. As an observer with a clear rooting interest, my belief all along was their external explanation was in deep contrast to their internal decision-making process. It's similar to politicians appearing on cable news programs to recite talking points that are the opposite of their core beliefs. It always amazes me how they can throw apparent BS to the public with such a straight face and then be so brazen as to fall on the sword of being "principled." Moving past my bipartisan rant, I'm convinced that Jerry, along with Filo, were loathe ever to sell their creation to the "Evil Empire." This was always more about *not selling to Microsoft* than it was about not selling at all. Jerry is hyper-competitive and he did believe he could turn the company's fortunes around, but if unsuccessful, the thought of Microsoft owning his creation was a non-starter.

If Yahoo!-eBay would have created a culture war, a Microsoft-Yahoo! merger would have set off the Tsar Bomba—the most powerful nuclear bomb ever detonated, with an explosive force of over fifty megatons and more than three thousand times more destructive than the nuclear bomb that was dropped on Hiroshima. (The Tsar Bomba was detonated as a test by the Russians on October 30, 1961.)

Jerry had an awful dilemma. Although the offer was very fair to Yahoo! shareholders, it would not only end Yahoo!'s time as an independent operating company, Yahoo! would likely be "Netscaped," never to be heard from again. Microsoft infamously crushed its competitors and sent their remains to "sleep with the fishes," but not just fish that live in shallow waters. Yahoo!'s remains would be sent to live in eternity with viperfish and daggertooth fish, each of which survive thirteen thousand feet down in the deepest and darkest parts of the ocean. Microsoft wasn't the type of company like a Time Warner, Procter & Gamble, or a General Electric, which were each set up more as a holding company for distinct brands. Microsoft *was* the brand, and there was no conceivable way to imagine that Ballmer would have kept the Yahoo! brand intact.

Although Microsoft hadn't released Bing, its engineers were working around the clock to develop a web-based search engine to compete with Google. Yahoo! had invested hundreds of millions of dollars in the development of Panama, which, although it didn't achieve its aspired-to success, was most certainly a well-designed functional search engine.

As a former Yahoo, I secretly hoped Jerry could thwart the hostile takeover. I had confidence in Jerry, so much so that when he ascended to CEO, I bought stock in Yahoo! for the first time since I had left the organization six years earlier. As a shareholder, I knew the Microsoft offer was a good deal, but my heart outweighed my pockets.

The obstacles for a transaction ultimately being completed went further than just Jerry. Google was determined to see the deal killed. Even though Microsoft had settled its antitrust violations with the government years earlier, as part of the agreement there was a probationary period that lasted until 2009; a federal court and technical experts were monitoring any activity that could result in further monopolistic behaviors. Google was well aware of the decree. David Drummond, Google's SVP and chief legal officer, summarized Google's concern that Microsoft ownership of Yahoo! could "exert the same sort of inappropriate and illegal influence over the internet that it did with the PC."

While Microsoft publicly acknowledged that they saw no obstacles to final regulatory approval, Google, along with its unlimited resources, planned to put up a dogged challenge to the merger. Even if the deal were ultimately greenlighted, Google had the potential to slow government approval to a crawl.

Google played more than a passive role to thwart the transaction. Eric Schmidt, Google CEO, was in contact with Jerry to determine if there was a partnership that the two rival companies could identify. Also, Google executives frantically worked to determine if there were other potential bidders for Yahoo!

Yahoo!'s board and bankers were also putting out feelers to fend off Microsoft. There were informal discussions with AT&T, Time Warner, Comcast, NewsCorp, and likely dozens of other potential corporations with the resources to bid against Microsoft. Options beyond a full acquisition were considered. Yahoo!, even with all its missteps, still owned several valuable products such as Yahoo! Finance and Yahoo! Sports. In the end, breaking up Yahoo!, which had frequently been discussed over the past decade, was determined not to be an option. A major media company senior executive told *The New York Times* on February 4, 2008, with regard to breaking up Yahoo! and selling its assets individually, "no one can get to a forty-four-billion-dollar price." Here was the biggest obstacle for Yahoo! and other potential bidders to overcome: Microsoft's unsolicited forty-four-billion-dollar bid was far and away the best offer. Ballmer wasn't joking around. He knew his offer was outstanding and even if other potential suitors appeared, no company had the resources to tangle with Microsoft in a bidding war, except possibly Google. With Bill Gates's retirement, Ballmer elevated to emperor without giving up the title of Darth Vader. The problem for the rest of the universe was the Jedi's "Force" was weak. The "Evil Empire" this time had arrived armed with Kryptonite against the one Jedi that could defeat them—Google. (And yes, I integrated Superman with Star Wars and referred to Google as a Jedi warrior, but sometimes analogies just aren't very clean!)

By 2008, Google was the clear winner in search, but Yahoo! was a solid number two. The combined entities would have screamed anti-trust so loudly that the Federal Trade Commission would have thrown a party complete with fireworks to stamp NO on to a Google-Yahoo! merger.

With all options denied, it was left up to Jerry and the board to reject Microsoft's offer. While Google had the potential to slow the transaction, it likely would have been eventually approved, if for no other reason than Google had secured such a large lead in search that a combined Microsoft-Yahoo! would have *still* been a distant second. And a winning case could have been made that it would have provided advertisers with another choice to spend their ad dollars.

On February 11, 2008, ten days after the initial offer, Yahoo! officially rejected Microsoft's bid. I was thrilled. Yahoo! would not be sold to Microsoft or broken up into pieces in an auction. Even with all the missteps, and there were many, deep down I was rooting for Yahoo! to succeed.

The problem was my vote meant a little less than zero. Yahoo! still had to answer to its largest shareholders. After all, the Microsoft offer of thirty-one dollars a share was not only more than fair, they had signaled they were willing to increase the price to close the deal, which they ultimately did. By May 3, Microsoft officially withdrew its offer. Less than two weeks later, on May 15, the first of several shareholder revolts led by activist investors would be initiated.

CHAPTER 17
The Activist Investor—Take 1

Carl Icahn, Queens-born and the Jewish son of an atheist Cantor (yes, you read that correctly), has been called the greatest corporate raider of all time. His claim for that dubious title was his successful 1985 takeover of Trans World Airlines. He sold the assets to cover the cost of the takeover and left TWA with hundreds of millions of dollars in debt after taking the company private. The iconic character Gordon Gekko, in the film *Wall Street*, is based partially on Carl Icahn. In the film's central business transaction, Gekko's aborted takeover of Blue Star Airlines was inspired by Icahn's successful takeover and then destruction of TWA. Unfortunately for the management and employees of TWA, Bud Fox, who was the architect of the Bluestar employee union-led rebellion, was just a character in the film. Of course, I'm sure in real life, no employee would feel comfortable having Charlie Sheen (who played Bud Fox) running their company. Sure, they would have tons of fun, but their fun would probably be paid for by raiding their pension fund.

Somewhere along the way the term "corporate raider" was softened to "activist investor." Although the title was changed, the job description remained the same. So "activist investor" Carl Icahn sent a letter to the Yahoo! board, which shared it with the public. In the letter, Icahn urged Yahoo! to accept Microsoft's offer of thirty-three dollars per share. (Before withdrawing its offer, Microsoft had increased the price from thirty-one dollars per share to thirty-three dollars per share.) At the time of the offer, Yahoo!'s stock price was $19.18. Although it had steadily climbed to over twenty-seven dollars by the time Icahn sent his letter to the board, the increase in share price was due entirely to anticipation of a Microsoft takeover and not because of increased value in the core Yahoo! business.

Icahn had acquired fifty-nine million shares and was seeking approval from the FTC to acquire another 2.5 billion dollars in Yahoo! shares. Also, other large shareholders were furious at Yahoo! for rejecting the Microsoft offer and were backing Icahn's revolt. With significant leverage, Carl Icahn nominated a slate of ten directors, including Mark Cuban who'd made his fortune selling his company Broadcast.com to Yahoo! If approved at the annual shareholder meeting, Icahn would have replaced the existing board, fired Jerry Yang, and re-opened discussions with Microsoft to purchase Yahoo! Unfortunately for Carl Icahn and fortunately for Yahoo! employees, after Microsoft withdrew from consideration, they indicated no interest in re-opening takeover discussions. However, the damage had been done and the two sides announced a settlement on July 21, 2008, confirming the Yahoo! board would increase its membership from ten to eleven, and Carl Icahn would control his seat and the appointment of two additional members of the Yahoo! board of directors.

Although Yahoo! entered into a settlement with Carl Icahn, there was no way to ignore that he'd inflicted significant damage. During the seven months that the company battled to thwart the takeover and investor uprisings, management was entirely distracted. Little effort to launch new products or improve existing products materialized. It is virtually impossible for any organization to spend over half a year being inattentive to its core business. And Yahoo! was just not any business; it was an iconic brand losing considerable ground to Google and Facebook. Losing seven months of activity is akin to losing seven years.

CHAPTER 18
Moral Pygmies

Unfortunately for Jerry, the Microsoft takeover and the Icahn investor uprising weren't the only non-core business-related issues during his tenure. On September 24, 1999, Yahoo! China officially launched. Jerry, a rock star in China, was on hand for the opening. In subsequent years, Yahoo! China was unable to generate much enthusiasm. Ironically, Yahoo! Japan, a joint venture with Japanese conglomerate Softbank, was Japan's leading internet business. In October 2005, after months of complex negotiations, Jerry engineered an agreement to invest one billion dollars and to transfer the assets of Yahoo! China into the burgeoning Chinese corporation Alibaba for a 40 percent stake in the pioneering e-commerce company.

For Yahoo!, this meant having to obey the strict laws of the communist nation. However, by giving up majority ownership, Yahoo! had minimal ability to pressure Alibaba to oppose the Chinese government on human rights issues—which completely countered the views of the United States. For Yahoo! China's operation, the phrase "May you live in interesting times," a traditional "Chinese curse" (though no evidence an actual Chinese source exists), rang true. If the holy grail in life is to possess all of the influence with none of the responsibility, then Yahoo! China had the misfortune of having all the responsibility with none of the influence. Or, as the nearest Chinese expression to the "Chinese curse" translates, "Better to be a dog in a peaceful time than to be a human in a warring period."

In November 2007, Jerry was forced to apologize to the House Foreign Relations Committee for Yahoo!'s role in the disclosure of journalist Shi Tao's dissemination of state secrets to the Chinese government. In 2004, the Chinese government ordered a complete media blackout regarding the fifteenth anniversary of the Tiananmen Square citizen protests. Tao forwarded the edict to a pro-democracy group using a Yahoo! email account, was subsequently arrested, and was sentenced to ten years in prison.

During the hearings, Congressman Tom Lantos of the United States House of Representatives, whose district includes Silicon Valley, pointedly scolded Jerry and Yahoo!'s general counsel Michael Callahan with the scathing remark, "While technologically and financially you are giants, morally you are pygmies." In addition to offending pygmies, he forced Yang and Callahan to stand and face Tao's mother, Gao Qin Shen, and "beg the forgiveness of the mother whose son is languishing behind bars."

I have no idea why Congressman Lantos would go so far as to offend pygmies; after all, that is akin to offending dwarfs, midgets, and little people, but the statement reverberated and was reported worldwide the following day. Also, it's a tad hypocritical for any member of Congress to refer to any individual offensively. It's possible that Congressman Lantos was just looking in the mirror of Congress when he made that statement. As an aside, I unequivocally believe that Jerry Yang is a man of high integrity and I personally feel offended by the comments of the members of the House Foreign Affairs Committee aimed at him, especially coming from a body filled to the brim with "moral pygmies." Yes, I apologize to pygmies.

On November 17, 2008, seventeen months after being hired to replace Terry Semel, Jerry Yang resigned as the third CEO of the company he'd founded thirteen years earlier.

In a recap of my running tally of Yahoo! CEOs, Yahoo!'s first CEO Tim Koogle's tenure was wildly fruitful for the first five years while he led the organization until his reign came crashing down with the dot-com bubble. Terry Semel, who also lasted five years, will be remembered mostly for losing search to Google and social to Facebook. Jerry, in his seventeen months as CEO, sadly will be remembered for his rebuff of Microsoft's hostile takeover, uprisings from activist investors, being slapped down by Congress, a falling stock price, and two rounds of painful layoffs. While it's clearly the understatement of the millennium to acknowledge that the board never anticipated Jerry's tenure would last only seventeen months after his elevation to CEO, no one could have predicted the outside forces that compromised his ability to revive the ailing company. On the day of Jerry's resignation, Yahoo! stock was trading at $10.63, a far cry from the thirty-three dollars offered by Microsoft.

CHAPTER 19
Carol Bartz and the Evil Empire— Take 2

On January 13, 2009, Carol Bartz was named the fourth CEO of Yahoo!, the third new leader in less than two years. If the opposite of a match made in heaven is a match made in hell, then this was a match made in burning volcanic lava at the bottom of the RXJ1347, which just happens to be the name of a cluster of galaxies five billion light years away with a temperature of three hundred million degrees Celsius, making it the hottest identified spot in the universe.

Carol Bartz had a successful career leading Autodesk, an engineering and architectural design software company. Also, she was well known in Silicon Valley, having been an elected member to the boards of such luminary Silicon Valley companies as Intel, Cisco, and others. If Koogle was mild mannered, Semel was elusive, and Jerry was friendly, then Carol Bartz was abrasive. She lobbed more "f-bombs" in one conversation than the entire stadium of fans at an Oakland Raiders football game.

From the start, she was the wrong fit. She was hired to manage a turnaround, but her tenure was memorable only due to significant cost cuts through layoffs (about 5 percent of the staff), shutting off products, outsourcing search to Microsoft, and activist investor uprisings. She was a hard-charging executive with very little vision for a company that was desperately searching for it. How little vision did she have? Her elevator pitch for Yahoo! was, "The place where people find relevant contextual information about things they care about." Huh? Who helped her come up with this brilliant explanation for Yahoo!? Al Gore? John Kerry? Mitt Romney? A thirteenth-century philosopher? Or some other Earth-shattering orator. Lincolnesque, it wasn't. Some poor speechwriter should be banned for life from writing speeches after allowing this to become Yahoo!'s introductory narrative.

Not only was there no product vision, there was product *implosion* of existing services, none more notable than search. In an August 7, 2009, interview with *The New York Times*, Bartz famously declared about Yahoo! that, "We have never been a search company." Somewhere I'm sure that Terry Semel, Jeff Weiner, Jerry Yang, David Filo, and the rest of the organization invested in Project Panama burned the newspaper in effigy in protest. In that same interview, she declared that having 20 percent of the search market was "very viable" and went on to state that with that number "we would be happy all day long." You are kidding…please tell me you're joking. Please? Double please?? Triple please???

To illustrate how disconnected to reality that statement was, on the day *The New York Times* interview was published, Yahoo!'s stock price was $14.62 while Google's was $227.61.

Nearly all of the revenue Google was generating was exclusive to paid search, while Yahoo! was still relying on a display advertising model that was one breath away from flatlining. While Semel had pretzeled himself between becoming a viable search company or a viable media company, Bartz was going to demonstrate the clarity that the market was craving and unequivocally announce that Yahoo! was not a search company. The problem was not just that this declaration extended Google's insurmountable lead even further; it was also that Bartz re-branded Yahoo!'s media segment as "relevant contextual information." Sports, News, and Finance at least glided off the tongue more effortlessly.

At the time Carol Bartz joined the company, the shareholders still had Yahoo! in a chokehold while they applied intense pressure either to sell parts of the business or the entire operation. Wall Street was not going to let go unless some meaningful transaction occurred. With the government strongly hinting that a partnership with Google would be denied, and Bartz's declaration that Yahoo!, after years of missteps, would be getting out of the search business, her only choice was to turn back to the "Evil Empire."

Steve Ballmer welcomed Carol Bartz with open arms. After all, his primary interest in Yahoo! the previous year was its search engine. Now Ballmer had the best of all worlds. He had control of the Yahoo! search engine for forty-four billion dollars less than he originally offered. My assumption of most people is that if they could have their wish come true for a forty-four-billion-dollar discount, then that would be considered a pretty good day.

On December 4, 2009, almost eleven months into Carol Bartz's tenure, Yahoo! and Microsoft announced its search partnership. The deal was very complicated. The following is my best attempt to simplify it. In the July 30, 2009, edition of *Search Engine Land*, search guru Danny Sullivan provides the simplest explanation I've come across in my research. According to Sullivan, the featured element of the deal is, "Yahoo is going to give up their technology and use Microsoft's."

Microsoft's Bing search engine was officially launched on June 3, 2009. For several years, Microsoft had been playing catch-up both to Yahoo! and Google. By the time of the partnership announcement, according to ComScore Inc., Google handled 65 percent of U.S. search traffic, Yahoo! covered 20 percent, and Bing had 8 percent. Yahoo!'s missteps and Microsoft's late entry into search resulted in Google also owning the dominant technology platform as well. However, Yahoo!'s earlier efforts with Project Panama produced a very capable search engine for users and advertisers.

Ironically, according to Sullivan, details of the partnership revealed that Bing's technology would be prominent, while parts of Yahoo! Search that would enhance the Bing product would be integrated. Yahoo! would maintain control of their user interface while each search result on Yahoo! would be revealed to be "Powered by Bing." On the surface, it seems contrary because, by most observers, Yahoo!'s search product development was more advanced than Bing. I know; I am both confusing and boring myself as well, but stick with me, this part will be over shortly and it's important. As a presidential candidate would say during a debate, "Just trust me."

The confounding agreement delineated that, while Microsoft owned technology responsibilities, Yahoo!'s sales force was responsible for monetizing search on Yahoo.com and selling ads on Microsoft's premium products. Yahoo!'s cut of the revenue, according to Sullivan's reporting, was 88 percent while the remaining 12 percent was apportioned to Microsoft.

Microsoft guaranteed Yahoo!, according to Sullivan, "Payments for the first eighteen months of the deal, a baseline of what it was earning before the deal starts," or approximately five hundred million dollars. The length of the agreement was for ten years during which time Yahoo! would provide Microsoft with an exclusive license to Yahoo! Search. The two companies also negotiated a "check-in" period after five years, at which time either party could revise the agreement or continue for the full ten years.

The last, most confusing, and ultimately the most important section of the partnership was Microsoft's agreement to a guaranteed "Revenue per Search," or RPS. Renowned search expert Danny Sullivan describes in the May 7, 2009 edition of *Marketing Land*, "Yahoo was promised that it would earn an amount for each search that happens that was somewhat competitive with Google." Sullivan continues by providing an example that if "Google was earning on average forty cents per search, then Yahoo might have been promised to get close to this with Microsoft's help. For example, maybe Microsoft promised that Yahoo might make twenty-five cents per search." The two companies never disclosed where the RPS was set.

Although it was a highly anticipated partnership, the markets were unmoved by this announcement. On Monday, December 7, 2009, Yahoo!'s stock opened at $29.78 on the news after closing on December 5, the day of the announcement, at $29.98. Microsoft closed on Friday at $15.19 and opened Monday at $15.36.

There is little question in my mind that Microsoft made out like a robber holding up a bank with no security and the vault wide open. Around June 16, 2008, approximately five weeks after Microsoft withdrew its offer of thirty-three dollars per share, a less audacious attempt was made by Ballmer to purchase the search assets of Yahoo! exclusively. According to Sullivan, Microsoft would have acquired Yahoo!'s search for one billion dollars and invested an additional eight billion dollars in Yahoo! at thirty-five dollars per share. Instead of any investment or full acquisition, Ballmer holding out resulted in Microsoft controlling Yahoo's search at no upfront cost. It was as if the New England Patriots traded Tom Brady to a non-contending NFL team for a seventh round draft choice. The difference is Carol Bartz is no Bill Belichick.

As a postscript, the Microsoft-Yahoo! partnership wasn't a complete failure, but neither was it a roaring success. If graded, a C- would likely suffice. According to my research, Microsoft never was able to achieve the revenue per search goals that were set (although never disclosed), allowing new Yahoo! CEO Marissa Mayer and her team to renegotiate the agreement with more favorable terms for Yahoo!, five years and five months after the announcement of the original arrangement (Steve Ballmer had resigned as CEO of Microsoft in February 2014). In March 2015, according to web measurement firm ComScore, Inc. Google had 64.4 percent of the U.S. search market, with Microsoft following at 20.1 percent, and Yahoo! in a distant third with 12.7 percent of the market. A reexamination of the search market revealed in 2009, Google's reach percentage was nearly identical, while Yahoo!'s and Microsoft's share of the market was inverted.

Also, in 2014, according to the market research firm eMarketer, Google owned 71.4 percent of the U.S. search ad revenue, Microsoft had 10.5 percent, and Yahoo! limped along with 5.7 percent.

Carol Bartz thus became the third Yahoo! CEO to bomb at implementing a search strategy. Tim Koogle chose not to pursue paid search, Terry Semel undervalued Google and was late to develop a true competitor, but Bartz simply threw up her arms and gave search away. After all, Yahoo! was just going to be "The place where people find relevant contextual information about things they care about." Whatever that means! (Jerry Yang, in his short tenure, it can be argued, at least attempted to save search by fighting Microsoft's offers.)

By 2011, it was becoming increasingly clear that not only had the value of Yahoo! declined, but there seemed no obvious path back to credibility. Internally, it must have felt as if volcanic lava were turning to rock, the trees were petrifying, and dinosaurs were fossilizing. Shareholder value was flat, and innovation was sparse. Talent was running out to join high-flying start-ups faster than politicians run toward a television camera. Also by 2011, the five primary segments of the internet had been established: search, social, commerce, video, and mobile. Without a leading position in any of those areas, there was no path toward market domination and ultimately corporate independence. The tragedy was, that from 2001–2011, Yahoo! was positioned to lead in all segments. By bumbling the Google, Facebook, and eBay deals, the company lost out on search, social, and commerce long before Carol Bartz was hired.

Carol Bartz's poor decision-making and limited vision put the nail in the coffin for search and effectively ensured Yahoo!'s eventual fate as a struggling organization doomed to be owned by a phone company and renamed "Oath." Ultimately, however, it was her inabilities to define the corporate mission, generate a positive corporate culture, inspire engineers, and develop shareholder confidence that led to her firing on September 6, 2011. The board had become so disillusioned with her performance that chairman Roy Bostock notified her by phone that she had been terminated. On January 13, 2009, the day of her hire, the stock closed at $12.10. Thirty-three months later, the day Carol Bartz was fired, Yahoo!'s stock price closed at $12.91.

In the years prior, Yahoo!'s inability to define itself as technology company or a media company resulted in *awful* stagnation. Ironically by 2011, sixteen years after Yahoo! incorporated as a business and eighteen years after the launch of Netscape—credited for blowing the doors off to the world of the internet—the industry was coalescing. The question that had haunted and perplexed Yahoo! for years was asked and answered worldwide. The mystery of the Sphinx had already been solved. To be a successful online consumer business, you needed to be both a media company *and* a technology company. Whereas Tim Koogle and Terry Semel struggled to figure that out, by the time of the Carol Bartz regime, that should have been as obvious as the burning seventy-seven story tower that she ran Yahoo! into when she gave up search.

Google, with the dominant search platform, had a deep, engineering-driven culture. Their programmers were encouraged to innovate and take risks even if they ultimately failed (example: Google Glass). The Google advertising platform and targeting capabilities were enhanced by its acquisition of Doubleclick, a company with no consumer familiarity but which possessed powerful technology tools that enabled Google to deliver ads across the entire online universe. Google-owned YouTube would arguably evolve into the most effective media service in the history of the internet. At the annual

YouTube Brandcast event on May 4, 2017, YouTube CEO Susan Wojcicki claimed that "more people 18-49 visit YouTube on mobile devices alone than any broadcast or cable TV network during primetime." Not to be lost in the audience figures, YouTube also represents over five billion dollars annually to Google's overall revenue. Google's secret sauce is that they leverage their technology platform to deliver a twenty-first-century media experience.

Probably no company has merged technology with media better than Disney. When legendary CEO Bob Iger was hired on March 13, 2005, one of his first orders of business was to reach out and form a partnership with Steve Jobs. Before Iger's promotion, Disney had become stagnant. Its animation business had gone into hibernation, their broadcast network ABC, after years as number one, had slid behind NBC and CBS, and Tomorrow Land at Walt Disney World screamed 1975. I always sensed they were still trying to sell color TVs as the future! Iger was intent on infusing technology into every division within Disney. With Jobs's assistance, he was the first media executive to embrace the iTunes delivery platform fully. He upgraded his retail shops to feel more like an Apple Store. He acquired technology innovator Pixar, which resulted in Jobs's election to the Disney board, and each subsequently released film increased the bar for innovating technology in animated films. His engineering teams developed a radio frequency identification wristband to improve the experience at the Disney theme parks, and most notably he embraced the internet with several significant acquisitions such as YouTube Channel Maker Studio and a hefty investment in MLB.com. A feature article from the December 29, 2014, issue of *Forbes* claimed of Iger, "In fact, he's the closest thing the company has to a central head of tech. And while there are many lessons to learn from the way he had run Disney over the past decade, this one is right up there: Not only do today's media companies need to start thinking like technology companies—their CEOs also need to start thinking like CTOs." [Chief Technology Officers]

It's ironic that in 2001, following AOL's titanic announcement of their merger with Time Warner, Wall Street suggested that Yahoo! acquire Disney in response. As of May 2017, Disney's value was 170 billion dollars more than Verizon paid to acquire Yahoo! So while Yahoo! struggled to define its strategy, as of the mid-2000s, companies were already beginning in earnest to embrace the intersection of media and technology. Not only was it shameful that Yahoo! failed to recognize this evolving trend, for the businesses that successfully did so, their shareholders were handsomely rewarded.

This was perhaps Carol Bartz's largest failing and by extent, the Yahoo! board's failure. They hired an executive to manage operations, cut costs, and streamline departments. That's a fine profile for managing a fifty-year-old property insurance company, but not for an internet icon that needed visionary leadership to turn around a sinking aircraft carrier in an ocean surrounded by wave runners and cigarette boats.

Bartz's other notable failure was her strained relationship with Alibaba founder and CEO Jack Ma. By 2011, it was becoming apparent that Jerry Yang's bold investment in October 2005 was paying off handsomely. In fact, to most observers, this was viewed as the greatest investment a U.S. company had ever made into a Chinese business.

In 2011, Yahoo!'s stock was languishing around twelve dollars. Their display advertising business was floundering, and they outsourced their search business to Microsoft. The company was still reeling from the aborted takeover attempt by Microsoft and analysts were starting to indicate that Yahoo!'s value was comprised entirely of its investments in Alibaba and Yahoo! Japan. By 2016, with turnaround efforts a failure, the Asian investments had grown so enormous that Yahoo!'s shoulders could no longer hold them, thus resulting in the sale to Verizon.

In May 2011, approximately four months before to Carol Bartz termination, it was announced that Alibaba had transferred ownership of its online payment system Alipay to Jack Ma and other investors without informing Yahoo!—its second largest investor after Softbank—of its intentions. The sheer brazenness of this act to make such a critical decision without Yahoo!'s knowledge was not just a slap in the face to Carol Bartz and the Yahoo! board, but it was more like a bat to the head. Ma demonstrated his total lack of regard for Bartz. To pour additional vodka in the pus-filled wound, Jerry Yang was a member of the Alibaba board of directors and he had failed to communicate the spin-off to Bartz. Following this episode, a third-grader could have predicted Bartz's ultimate fate.

CHAPTER 20
More Turmoil at the Top and Activist Investors—Take 2

With Bartz gone, CFO Tim Morse was elevated to interim CEO while the board of directors commenced a search for the fifth CEO (excluding Morse) in Yahoo!'s sixteen-year existence. Morse's reputation was as a cost cutter. As CFO, he certainly was not a visionary, nor was that the role expected of him. However, it created a problem from September 2011, when Bartz was fired, through January 2012, when Scott Thompson became the new CEO. Another quarter went by with turmoil, no innovation, fears of layoffs, and an exodus of top talent.

January to February 2012 were perhaps two of the most tumultuous months in Yahoo!'s turbulent history. Four major events occurred that both paralyzed the business and sent shock waves through the industry. On January 4, 2012, Scott Thompson was hired as the fifth CEO of Yahoo! On January 17, 2012, Jerry Yang unceremoniously resigned from the company he had co-founded in his apartment at Stanford University. On February 7, 2012, longtime Chairman Roy Bostock and four additional board members announced they would not stand for re-election, and on February 13, 2012, shareholder and activist investor Daniel Loeb, of hedge fund Third Point LLC, launched the second major proxy fight in less than four years.

Scott Thompson had previously been president of the eBay subsidiary, PayPal. Before his ascension to president, Thompson served as chief technology officer for the online payment pioneer. Unlike Semel, who had no technology experience, or Bartz, who had no background running a consumer-directed business, Thompson had success in both segments. Notwithstanding that, his tenure lasted as long as an on-time flight from San Francisco to Los Angeles, without, of course, the delay required to violently drag a passenger off a flight who was peacefully in his seat bothering no one as what occurred during a United Airlines flight on April 9, 2017.

Thompson's problem wasn't his experience. It was his lack of political skills and his inability to tell the truth. I know how backward that statement is. Usually, an individual's inability to tell the truth would make them a fine politician.

On January 17, Jerry Yang announced to the board of directors that he would be stepping down entirely from Yahoo! Although there are opposing views whether he was pushed out or left on his own, it was apparent that as the stock price floundered, shareholders were unforgiving about his refusal to accept Microsoft's offer in 2008. Kara Swisher, in her January 17, 2012, article on Jerry's resignation in the digital publication *AllThingsD*, reported that Yang's decision to leave was his own and the suddenness of it surprised all Yahoo! executives. In her report, she stated, "Yes, he jumped, even though being pushed was surely looking on the horizon ahead." She went on to claim that activist investor Dan Loeb and other shareholders were in favor of Jerry's removal because of their belief that he stood in the way of the company being sold. Also, Swisher revealed that "execs at both China's Alibaba Group and Japan's Softbank pointed to a too-strong influence of Yang in the deal as a possible stumbling block" to Yahoo! selling off parts of its stakes in the Asian assets. She also indicated that Jerry's inclusion as a board member of both Alibaba and Yahoo! Japan would be a conflict. (Along with resigning from

Yahoo!, Yang also resigned his seats as the Yahoo! appointee both to Alibaba's and Yahoo! Japan's board of directors.)

When I read this news, I admit I had a knot in my stomach. Jerry Yang was instrumental in my being hired at Yahoo!, which was the opportunity of a lifetime. Although we hadn't spoken in years, I had a fondness for our earlier friendship. He brought together brilliant dreamers who pioneered the internet. His approachability and kindness created a culture where each morning every Yahoo! walked into the purple-and-yellow "playground of opportunities." The training in the early years resulted in a Who's Who of legendary Yahoo! alumni, and he represented his creation and our soul with dignity and integrity for sixteen years.

On the flip side, Jerry adamantly refused to sell his company, misjudged the competition, and, equally as damaging, recruited CEOs who lacked vision. However, when I read the news that Jerry had walked away from the only adult job he'd ever held, I wasn't reflecting on the negative. Instead, I focused on an individual who had been raised fatherless who came to a foreign country as a child, without knowing the language, and who succeeded in changing the world. Jerry's creation pioneered how all information is discovered and delivered. Jerry took our vast world and made it smaller. It's an understatement to call that quite a legacy, and I'm proud to have known him.

As a postscript, Jerry was reappointed to the board of directors of Alibaba and is now one of the most successful early stage technology investors in the Silicon Valley.

With Jerry removing himself from any role with Yahoo!, more disarray ensued. On February 7, 2012 Roy Bostock announced his intentions to retire from the board of directors. Wall Street often criticized Bostock as being ineffective during his tenure as chairman, and along with Jerry, he was personally blamed for the Microsoft debacle and the hiring of Carol Bartz.

During his tenure, the stock prices remained flat for years. He provided minimal vision and was at the center of the anger of investor revolts. He steadfastly refused to sell or breakup Yahoo!, and he placed too much faith for too long a period in CEOs who failed to deliver shareholder value. While his and the resignations of four other directors pleased Daniel Loeb and his Third Point investors, it wasn't enough. Loeb wanted full control of the four new additions to the board and submitted his own slate naming three directors and himself.

Daniel Loeb spent early life in Santa Monica, California, where he grew up deeply associated with Mattel, the toy company that his great aunt co-founded. She was also the creator of the Barbie doll. From a very young age, he had a fascination with the stock market. In 1995, after banking stints at Jeffries LLC and Citibank, he launched Third Point. In the twenty-two years of Third Point's existence, the hedge fund has grown to be a powerhouse. In 2017 the firm's assets under management (AUM) total more than seventeen billion dollars. By 2012, Third Point had acquired Yahoo! stock totaling 5.8 percent ownership of the company.

Unlike Carl Icahn—whose strategy was to leverage his position as a large shareholder to force Yahoo! to sell either as a whole or as parts—Loeb had different goals. Although he was not the CEO, an argument can be made that he was perhaps the most visionary figure in the company in the past twelve years. In Yahoo!'s global audience, which numbered more than a billion strong, Loeb saw the potential to unlock value. He was direct in displaying his disgust with Yahoo!'s executive merry-go-round, even opining that Yahoo! "had one of the most horrendous management teams" he'd ever seen. If Loeb didn't say it in public, he must have surely made the same statement about the board of directors at least privately. In his first call with Roy Bostock, he demanded Roy resign immediately. Ironically, his attacks on Yahoo! management began before Scott Thompson was hired. Once Thompson

WE WERE YAHOO!

was named CEO, Loeb immediately worked to undermine his tenure. He launched a crusade to make Thompson's career at Yahoo! the shortest reign since the College of Cardinals chose John Paul I. (Pope Paul I died September 28, 1978, only thirty-three days after being elevated to be the leader of the Catholic Church.)

Loeb was intent to shake up Yahoo! and to do so he needed a CEO that he supported. Acting like a private investigator finding dirt on a cheating spouse, he poured through Scott Thompson's entire adult history before uncovering a seemingly benign detail in his corporate bio.

Thompson had attended Stonehill College in Easton, Massachusetts. While it was a decent school, MIT and Stanford it wasn't. It wasn't exactly distinguished for graduating world-changing innovators or holding joint degree programs with Harvard or Tufts. In fact, the most notable alumni are a handful of Massachusetts state legislators, not to be confused with Thomas Jefferson or John Kennedy. As an aside, I grew up fifteen minutes from the school and the only memory I have of it was our high school soccer team holding an annual scrimmage with their varsity team.

According to Thompson's bio, he claimed to have a bachelor's degree in accounting and computer science, when, in fact, he only held a degree in accounting. Now, if he were a politician, he likely would have been promoted for fabricating his résumé, but as the CEO of a high-profile, publicly traded company with irate shareholders aiming a Smith & Wesson .44 magnum revolver at his head, this meant doom. Loeb, as astute in public relations as he was as an investor, had his Dirty Harry weapon of choice. And the gun was unlocked and loaded.

Of all of the hiring considerations for selecting Scott Thompson, one must have been the recognition that previous CEOs lack of technology experience had resulted in failures. Having technology experts leading an organization was comforting to an engineering-driven culture. While Thompson had technical know-how as a former chief technology officer at PayPal, his lack of a computer science degree undermined his credibility internally, and his deceit and inability to be forthright about it undermined the trust a CEO needs to have with the shareholders.

On May 3, 2012, Loeb revealed the impropriety to the board and subsequently released a letter to the media. The result was a five-alarm inferno internally at Yahoo! and externally throughout the market. After almost two weeks of deflecting his role in "résumé-gate" (at one point even placing blame on the highly regarded recruiting firm Heidrick and Struggles, which forcefully denied any participation in the oversight), Thompson resigned. On May 13, 2012, he informed the board of his desire to step away from running Yahoo!, indicating that he was being treated for thyroid cancer. The board reeled from the perceived ineptness and on the same day announced that chairman Roy Bostock and three other directors would step down immediately, rather than wait for the annual shareholder meeting. The board also voted to end the proxy crisis with Loeb. They appointed him to the board along with two additional members of his slate of candidates.

Thomson's five months were more than just a distraction. It was chaotic. Nicholas Carlson, author of *Marissa Mayer and the Fight to Save Yahoo!*, revealed in his book that Scott Thompson had planned to cut approximately one-third of the fifteen-thousand-member workforce. While industry watchers and analysts were demanding a massive staff reduction, according to Carlson, Thompson's plan was to cut entire divisions rather than poorly performing individual performers.

On May 14, 2012, the day after Scott Thompson announced to the board his intent to resign, employee research firm Glassdoor published an analysis, reporting that his internal approval rating was a paltry 31 percent, down from a high of 85 percent on the day he had been hired five months earlier.

From June 2007 until July 2012—for over five full years—Yahoo! was in turmoil. The missteps of the Koogle and Semel years were in actuality small seeds that grew into a terrifying forest, with witches, flying monkeys, and talking trees with frighteningly deep voices and branches that reached out to throw any intruders from their horses.

The stock price was depressed, activist investors were activating, and highly skilled employees were racing out the door. New ideas were stagnant, M&A was minimal, and competitors such as Google and Facebook were achieving unprecedented levels of growth. The bureaucracy was choking, the culture had eroded, and the brand was exhausted.

However, as ominous as it has all been described, the company had a heart and a soul, even if it lacked a spirit. Over one billion individuals across the globe still used Yahoo! products and services. The Finance, Sports, News, and Email properties were ingrained into the daily habits of people worldwide. Yahoo! still attracted enough user traffic to be ranked among the top five most-visited websites in the world. While observers were all eulogizing the company, and calling for it to be sold entirely or spun out as individual pieces, Daniel Loeb concluded there was significant value left in the organization. And the key to unlocking that value was to recruit a CEO who was youthful, energetic, and could instill a culture of rebirth and reinvention. A leader who could control costs, yet not be afraid to take risks. A chief revered both by engineers and advertisers, and an individual with recognizable accomplishments as an innovator.

CHAPTER 21
Marissa Mayer: Hope and Disappointment

On July 16, 2012, the company announced that longtime Google executive Marissa Mayer would be new chief executive officer of Yahoo! Inc. The thirty-seven-year old Mayer beat out Yahoo! interim CEO Ross Levinsohn to become one of the youngest CEOs of a publicly traded company. Her hiring, driven in large part by Daniel Loeb, was met with universal applause, most notably inside the walls of Yahoo!

Spoiler alert: This book is not about Marissa Mayer or a retelling of the story of her Yahoo! tenure. Her career as the CEO has been followed, analyzed, overanalyzed, and over-overanalyzed. Journalists have knocked one another down to file reports. There were periods where it seemed as if she was covered more ardently than Hillary Clinton and Princess Diana combined. (Just a little exaggeration for dramatic purposes.) There is certainly enough documentation, columns, stories, and opinions written about her that a mere book wouldn't do the story justice. It probably requires a full twenty-two-episode season to give a full account of her time leading Yahoo! Perhaps Terry Semel and Lloyd Braun can produce that series.

This is also not about laziness on my part or not wanting to research and read the thousands of pages that detail the years in which she led the company. I assure you, as an early employee, I had such a strong rooting interest in the business that there were very few articles detailing her reign that I haven't already pored over.

While critical moments and elements of her tenure will be examined, Marissa Mayer was not the reason Yahoo! was doomed to survive as an independent organization. The missteps of her predecessors predestined the final outcome. Marissa Mayer didn't choose to reject paid search. Marissa Mayer didn't choose to reject the acquisition of Google early on and then misjudge its value a few years later. Marissa Mayer didn't overplay her hand and allow Facebook to walk away from the negotiating table. Marissa Mayer didn't implode the merger with eBay. Marissa Mayer didn't devalue Flickr. Marissa Mayer didn't hand search over to Microsoft. Marissa Mayer didn't turn down a forty-four billion dollar offer from Microsoft. Marissa Mayer didn't neglect the burgeoning evolution of the mobile platform. Marissa Mayer didn't allow Google to acquire YouTube, Android, and Doubleclick and then perfectly execute the integration of their products and services.

Long before Marissa Mayer, Yahoo! had already been soundly defeated in search, social, commerce, video, mobile, and advertising. Mayer was hired to turn around the stale business by competing in the categories that had crystallized as the primary groupings for success in the online world. The challenge was far beyond daunting. In the approximate twenty-five-year history of the consumer internet industry, there isn't an example of a dot-com turnaround. Of all the internet businesses during that period, they break out quite logically into four distinct categories throughout its young industry existence:

- "Crashers." These companies enjoyed many days and nights of self-indulgence. Like David, King of Israel, who forgot that God's grace was responsible for his success, these businesses ignored that their triumphs were a result of irrational exuberance and not a solid business model with profitable quarter-over-quarter revenue growth. Companies such as The Globe.com, Pets.com, eToys, Webvan, NetZero, boo.com, Flooz, Kozmos, and countless others were ordained to cease existing.

- "Swallowed." Whereas, after a period of self-reflection, the whale released Jonah. Companies such as Infoseek, Lycos, Broadcast.com, and Geo-cities were swallowed by more powerful whales and their brands were destined to be lost for eternity.

- "Survivors." These companies were condemned to wander in the desert for forty years. Fortunately (or unfortunately), they were rescued while nearing death from years of neglect and missteps. Yahoo! and AOL fit perfectly into this category.

- "Thrivers." These are the companies whose deep souls and risk-taking spirit have assured them entrance into the Promised Land. The (ten) others were driven by the sins of emotion and fated to die by plagues. Only Joshua and Caleb acted with conviction and bravery. Google, Amazon, and Facebook lead the "Thrivers."

Turnaround efforts failed for the Crashers and the Swallowed. The Survivors were rescued but were left beaten and bruised with permanent scars. The Thrivers grew and stomped on the Crashers, devoured the Swallowed, and happily, let a phone company acquire the Survivors.

In the history of technology, there have been several examples of successful turnarounds of technology and media organizations. Most notable is Apple, which was flirting with bankruptcy before Steve Jobs returned to resuscitate the flailing computer maker. IBM was somewhere between the age of dirt and dinosaurs and every bit acting its age before Lou Gerstner was hired in April 1993. Disney had seen its prized animation division wallow with forgettable titles such as *Home on the Range, The Emperor's New Groove,* and *Treasure Planet* before Bob Iger's rise to CEO. While it was certain that the fortune of those businesses turned upward due in large parts to the vision of these legendary leaders, just as crucial was the sector they were competing in allowed for a turnaround to take place. That just simply wasn't the case for Marissa Mayer at Yahoo!

Apple is a hardware company. Their most significant invention, the iPhone, can be touched. Yahoo! existed only on a screen. Its ability to invent the "next great thing" was minimal. Technology moves at such speeds that stagnation over a few months, let alone five years, makes it nearly impossible ever to catch up. Even though the internet is just space, there is no wormhole to make up light years in travel. A "screen only" company that has lost in search, social, commerce, video, mobile, and advertising has nowhere to advance without transporting itself off the "screen" and into the world of touch and feel. Yahoo! was never equipped to do this. In fact, it was merely an extension of traditional media. Their second-tier position precluded them from the developing IoT products, wearable technology, smart speakers, drones, virtual and augmented reality headsets, self-driving vehicles, robots, and other innovations that companies with bottomless pools of capital could afford. While all of these technologies are the promise for the future, with few exceptions, they are cost-prohibitive and currently highly unprofitable. Companies like Apple (along with Google, Facebook, Microsoft, and some others) have the capital to invest in projects that will take years to materialize as financial juggernauts for the parent organization.

In the 1980s, IBM had lost its way. It was bloated and bureaucratic. It was investing billions of dollars in failed products (OS/2 Operating System is a prime example), plants, and laboratories. Departments were competing against one another rather than cooperating on behalf of customers. Gerstner changed the entire organizational culture by closing useless plants, selling assets, and shutting down failed products. He transformed the company away from just selling hardware and software to offering consulting services for businesses to integrate and modernize entire complex enterprises. The conversion to a consulting company positioned IBM to upsell their customers with the IBM products.

Again, as successful as Lou Gerstner was in turning around IBM by transforming the entire company's strategy, there was no similar model for Marissa Mayer to emulate. Integrating entire modern systems is enormously costly and highly complex. Purchasing advertising on search engines is inexpensive and a simple task. It does not require dozens of team members to be incorporated into the site for years at a time. The difference between purchasing online advertising and integrating complex enterprise systems is akin to the size of the gap between learning grade school long division versus PhD-level advanced calculus. Mayer was well aware of the divide, having graduated from Stanford with a BS in symbolic systems and MS in computer science (see Scott Thompson, now that's a real computer science degree) with specialties in artificial intelligence.

Years had passed since Disney's animation division had produced a bona fide hit. One of the directives of the Disney board (and particularly Roy E. Disney) was for newly minted CEO Bob Iger to revitalize the flagging Disney animation unit. Iger oversaw a renaissance at Disney Animation, which was removed from its corporate sister Pixar, with the release of the hit film *Frozen* in November 2013. The movie was such a success that it spawned an ice show, rides at the theme park, sequels, massive merchandise revenue, and a Broadway show. Parents of crazed younger fans have referred to the film and its music as crack for ten-year-olds. Like Apple and IBM before, it was a great achievement for Disney, but it did not offer any lessons for Marissa Mayer to mimic within Yahoo! Not to undervalue the difficulty of creating a great movie, but it's easier to develop a great film than it is to build a search engine to overtake Google or a social network to overtake Facebook.

At some point, a business clearly hits an inflection point, and it's right for the stakeholders to squeeze out any remaining value. At Yahoo!, that moment was arguably between the Carol Bartz and Scott Thompson debacles. However, leaders within the organization weren't prepared to grasp the ultimate outcome and chose to give the company one last shot. Thus, Marissa Mayer not only had the challenge of growing Yahoo! but first she needed to save it.

Usually, my to-do list is comprised of picking up the dry cleaning, some food shopping, a few conference calls, helping kids with homework, paying some bills, and other household or basic work chores. Marissa Mayer's checklist was just a little bit more demanding. Her tasks included rebuilding a positive corporate culture, regaining the search service from Microsoft, developing a social network business, converting Yahoo!'s products to a mobile platform, launching original programming, modernizing the stale look and feel of Yahoo! products, create new killer products, acquire strategic businesses, develop innovative ad programs to reclaim a larger share of online advertising budgets, please the new board of directors, improve the stock price, unlock the value of the Alibaba stake, and hold off those pesky activist investors. Outside of the occupier of the big oval office in the city tucked between Maryland and Virginia, it's hard to visualize anyone on earth with a more formidable list of challenges.

In order to execute the plan, Mayer first needed to confront the pessimistic corporate culture that had been expanding throughout all the turbulence of the past five years. The staff desperately needed a recharge. Years of layoffs, threats of massive staff reduction, talent drains, executive turmoil, growing bureaucracies, unfavorable press, sluggish stock price, and a loss of corporate purpose all contributed to a culture of unhappiness and unease.

Fortunately, she was welcomed as a savior—and unlike Scott Thompson—Marissa Mayer, with the full support of Daniel Loeb and the rest of the board, was going to have a honeymoon. On her first day, she was fêted with praise and hearty welcomes, with some employees displaying signs with the word "Hope." Wall Street endorsed the move, with the stock closing 2 percent higher at $15.98 after the news was announced.

The media reacted positively. The headline of the July 16, 2012, editorial in *Forbes* by contributor Dorie Clark read "3 Reasons Why Marissa Mayer's Hiring is a Huge Win for Yahoo." The opinion cited Mayer's hiring as the fresh start the company needs, Mayer's show of faith in Yahoo! after a decorated career at Google, and her knowledge of the business as the cause for a positive outlook. Dan Frommer, current editor in chief at *Re/code* wrote about Mayer's hiring in *Readwrite*, a web-based news and technology publication, on July 16, 2012, "This is a great move for Yahoo, which has stewed in mediocrity for years."

On a side note, post-Yahoo! I moved to Florida and was elected to the Florida State Senate. One of my great fortunes in life is to represent one of the largest populations of senior citizens in America. These amazing individuals include WWII veterans, former prisoners of war, and Holocaust survivors. They were born during the Great Depression, most never attended college but their children graduated from Ivy League schools and were doctors, lawyers, and bankers. From this population, I learned one undeniable truth. Free food is the most direct line to the human heart. During my campaign, my staff and I held town hall meetings in senior communities. If we served free coffee, we had a guaranteed crowd of one hundred attendees. If we provided bagels, we were guaranteed two hundred attendees, and if we offered lox and nova, the Holy Grail of the Brooklyn transplants, we'd be trampled by hundreds of seniors before we had a chance to unwrap the food. I refrain from reporting on the time we offered brisket sandwiches. The carnage was devastating.

As one of Marissa Mayer's first acts as CEO, she declared that all food served at Yahoo! would be free. Instantly, URL's (the name of the main company cafeteria) filled with smiling Yahoos. One employee who'd resigned during the Carol Bartz era to work for Google returned to Yahoo! when Mayer arrived demonstrated the 180-degree change in morale. It was more than free food, but other perks, such as free iPhones, and Friday all-hand meetings where exciting announcements were released and exceptional employees were recognized in front of their peers. She promoted team building and transparency that are the central aspects required to establish a flourishing engineering culture.

Within a few months of her hire, against the wishes of the board, industry observers, and analysts, Mayer announced to the staff that there would be no company-wide layoffs. All of Yahoo! cheered. She reinvigorated the entire organization, and for the first time in years, résumés of top talent streamed *into* Yahoo! rather than out. Unfortunately, the goodwill chips she was collecting were about to peak.

For all the fanfare of Mayer's arrival, it was quickly apparent that Facebook and Google weren't walking back through the door. But Mayer's chamber wasn't yet empty. Although total revenues in her first year as CEO mostly remained flat or even slightly lower, Yahoo! had amassed a significant cash position a few weeks before her arrival. Alibaba, the Chinese e-commerce portal, of which Yahoo! had purchased 40 percent in 2005, had grown into a juggernaut. While still a private company, some analysts reported that the online retailer was larger than Amazon and eBay combined. In May 2012, the board voted to sell half of its stake in Alibaba for 7.1 billion dollars. A few months later, the board announced that half of the proceeds would be given back to the shareholders, still leaving Mayer with considerable resources for a splashy acquisition.

CHAPTER 22
It Was Just Too Late

Marissa Mayer acquired over fifty companies during her five-year tenure with Yahoo!, but none larger and more exciting than the blogging social network Tumblr. On May 19, 2013, Yahoo! announced a cash deal to purchase the six-year-old company for 1.1 billion dollars. Yahoo! had a long history of screwing up acquisitions, from GeoCities to Broadcast.com to Flickr and dozens of others. It was so refreshing to hear Mayer announce that she "promises not to screw it up." In fact, she compared it to Google acquiring YouTube and eBay acquiring PayPal, two subsidiaries where it was invisible to the user who owned the businesses. From the moment I heard internal signals that the culture of the company was turning positive, I began slowly purchasing Yahoo! stock again. The day Yahoo! acquired Tumblr, I bought several thousand additional shares.

By 2013, Google had cemented that it was not just winning search or for that matter even dominating it, they were monopolizing it. According to digital performance solutions company NetMarketShare (warning, boring numbers ahead), Google share of the desktop search market was 77.46 percent and 87.59 percent of the mobile search market. Yahoo!, in contrast, had a measly 7.3 percent of the desktop search market and only 6.23 percent of the mobile search traffic. I should acknowledge as someone who spent years in the online marketing space that I've never heard the term "digital performance solutions company." In my very simple mind, I assume that just means *Mad Men* for the internet. That's so much more exciting.

Google is able to own the entire category because search isn't conducive to niche competition and it's much less generational than social networks. I admit I am fascinated by Sherpas. They are the most interesting ethnic group in the world. They survive in the highest altitudes on earth, have superhuman strength, and are the best travel guides on the planet. What's not to love? I've read books, read hundreds of articles, watched countless hours of video, yet, if I am searching for information, I start with Google. I don't have a Sherpa-only search engine. My assumption is that Kleiner Perkins or Sequoia Capital isn't falling over themselves to fund a Sherpa search engine business plan. (Note, I admitted early in this book it would provide lessons on what not to do in business, and I strongly recommend taking my advice on not wasting a single cent on building a Sherpa search engine. You'll be stuck with it. It has no clear exit strategy.)

Social networks are vastly different. Although Facebook was the clear leader, with 1.8 billion active members (as of January 2017) connected throughout the world and almost eight billion dollars in annual revenue, other social network websites and apps have launched successfully. Some, like Instagram, have been acquired by Facebook, and others, like Snapchat, have chosen an independent route. I use Facebook and Twitter. My kids use Snapchat and Instagram to connect with their friends. Pinterest targets women, while LinkedIn targets professionals. Most of the population have accounts on multiple social network channels. Google, in contrast, controls 95 percent of the mobile search traffic as of 2016 according to online researcher NetMarketShare. And not to be forgotten, I can connect with the billions of other Sherpa enthusiasts across the world on any of these platforms.

Search was out of reach, for sure, but acquiring Tumblr to aggressively push Yahoo! into the social network world made sense. Had Yahoo! acquired

Facebook it would have remade the company, had Yahoo!'s efforts to purchase YouTube been successful, it would have remade the company, had Yahoo! not botched Flickr, it could have remade the company. Marissa Mayer was intent not to follow in the same footsteps of the mistakes of the "Ghost of Terry Semel Christmas Past."

For all its promise, the social blogging network had one Achilles' heel that Marissa Mayer overlooked when she made her decision to go through with the acquisition. Tumblr had no significant track record of generating revenue. Mayer was under pressure to modernize Yahoo! and acquire or develop more products targeted at younger audiences. Unfortunately, for the team at Tumblr, she was under even greater pressure to increase revenue and profits for Yahoo! that had languished for years.

Forbes contributing writer Peter Cohan slammed the deal in a May 20, 2013, article. In the piece, Cohan cites, "Tumblr's reliance on interactive advertising versus Yahoo's display advertising as making the deal less attractable; Tumblr's inability to boost Yahoo's online advertising market share and the possibility of consumers bolting Tumblr because of its affiliation with an old stale 1990's business as predictions for the deal failing." In the article, Cohan indicates that Yahoo! vastly overpaid and that Google, Microsoft, and Facebook had all turned down the deal.

Cohan's predictions were mostly accurate. Tumblr's audience did not flee the service. As of January 2017, research firm Statista ranked Tumblr as the eighth largest social network site in the world with 550 million users, or an increase from 134 million users in 2013 when Yahoo! acquired the company. (Figures according to Quantcast and reported in *The New York Times* on May 19, 2013.)

Cohan was like Nostradamus with the rest of his prophecy. Marissa Mayer was intent on aggressively marketing Tumblr to advertisers, yet there were huge inherent obstacles that proved impossible to overcome. The first was that much of Tumblr's content was adult (i.e., porn) which is classified as "NSFW" (not safe for work). It's also safe to assume that Procter & Gamble was not rushing to spend its ad dollars on the Tumblr platform.

The second challenge, according to Cohan, was for marketers to be able to advertise on Tumblr, permission was required by the owner of the blog, which was not Tumblr. If permission was granted, then revenue was shared with the publisher. (Ironically, Facebook pioneered user-generated and user-approved display advertising and has proven it is a very lucrative ad model.)

At the time of the acquisition, Tumblr reported thirteen million dollars annually in advertising. Marissa Mayer imposed a first-year goal of one hundred million dollars toward Tumblr.

The revenue numbers for Tumblr were never released, but Yahoo! and industry statements indicate that Tumblr has underperformed for the company. Marissa Mayer's declaration was fractured when she did interfere in Tumblr's business. At one point, both sales teams were merged, although Yahoo! eventually realized that Tumblr would have greater opportunity to scale with an independent sales force, so it disjoined the two after the experiment had failed. By February 2016, Yahoo! wrote down the value of Tumblr for 230 million dollars, and in July of that year, wrote down another 482 million dollars. By mid-2016, more than half the acquisition had been written down, and analysts believed its value to be zero.

I don't blame Marissa Mayer for acquiring Tumblr. She was inserted into the game late in the fourth quarter as quarterback, with her team down by a score of 621 – 21. (I will allot them 21 points because let's not forget Yahoo! had a pretty effective first quarter.) Tumblr had warts for sure, but Mayer still chose to throw the ball downfield and took a shot at a wide open young receiver with vast All Pro potential, and the receiver dropped the ball. Perhaps the receiver should have stayed another year in college or perhaps the team could have better prepared the young receiver, but neither the quarterback nor the receiver were the causes of the six hundred point differential. Most independent observers did recognize the blowout and begged for the team to forfeit, but rather than quit they just came back, again and again, to have their heads blown in. As has sadly been proven, loads of shots to the heads cause early Alzheimer's and other degenerative brain disorders.

The acquisition of Tumblr also demonstrated Mayer wasn't afraid to take risks. At this point in Yahoo!'s lifecycle, risks needed to be taken, but it also validated that had she been CEO of Yahoo! during the Terry Semel years, she would not have undervalued Facebook. And if she had been the CEO during the Carol Bartz years, she would not have handed search to Microsoft. I do believe if Mayer were CEO during the Jerry Yang tenure, her competitiveness would have also led her to turn down the Microsoft forty-four-billion-dollar offer, but clearly, I am just assuming. (I am very confident, however, that Mayer wouldn't have gone to Stonehill College in Easton, Massachusetts, to receive a computer science degree.)

Mayer spent another one billion dollars to acquire at least fifty small companies. None were noteworthy, and her history as a dealmaker is certainly underwhelming. However, none of those businesses was ever going to turn out to be Google, Facebook, eBay, or YouTube; all real acquisition opportunities were ones the previous regimes botched.

Along with aggressively pushing Yahoo! to compete in the social network arena, Mayer wasn't ready to give up on search. While this may have been considered a fool's run with Google's commanding lead, there was no doubt that even minimal success would improve Yahoo!'s diminishing revenue. From the moment Mayer was hired, she wanted to shelve the 2009 deal with Microsoft. Fortunately for Yahoo!, the original agreement was only for desktop-based search. Mobile was far outpacing desktop as the preferred platform, and a new unit, called Native Ads, was rapidly advancing.

Native advertising is a kind of advertising created mostly for online campaigns that's embedded in the content of the website. It's the ads that appear on your Facebook newsfeed or integrated within a traditional content platform. It is clearly identified as advertising, but it maintains the same look and feel as the editorial content. The most visible form of native ads is paid search.

It's ironic and sad that much of Yahoo!'s struggles link directly to the decisions of the original product teams at Yahoo! that were intent on keeping an electric wall with a 25.5 megavolt barrier separating advertising and content (a megavolt is one million volts and 25.5 megawatts is the highest voltage ever produced). Almost twenty years later, the voltage has not only flamed out with the wall torn down, but virtually no user backlash has resulted. If Yahoo! had jumped off the cliff back in 2000, the parachute would have opened. Online advertising, both desktop and later mobile, evolved in the exact opposite manner against which Yahoo! so tightly guarded itself. Rather than embrace new rules for a new medium, they just extended the rules of traditional media onto a screen. The results were devastating.

To her credit, Marissa Mayer embraced native advertising. In February 2014, Yahoo! introduced its new advertising platform called Gemini. While not nearly as audacious a project as Panama, it accomplished a defined clear goal. Still choked by the Microsoft deal negotiated by Carol Bartz and Steve Ballmer, Gemini allowed Yahoo! the ability to compete both in native advertising and paid search on mobile devices, in which they were free and clear of the original agreement to do. In April 2015, six years after "gifting" their desktop search to Microsoft, the two companies announced a renegotiated agreement, allowing Yahoo! now to carry its ads on 49 percent of its desktop search traffic. In the original agreement, Yahoo! had to carry Bing ads on 100 percent of their search result pages. The Gemini system provided Yahoo! with a platform to serve ads across desktop search immediately. Without the new system in place, Yahoo! would have had to develop an entirely new platform, which would have set them back another eighteen months.

Unfortunately for Marissa Mayer and Yahoo!, by the time they had modernized their search platform and renegotiated the agreement with Microsoft, it was too late. In reality, it was probably too late in 2001, when Google embraced paid search. It was probably too late again when the Panama Project was delayed and its results fell well below expectations. And it was definitely too late six years after giving its search service away.

Google's share of the market was too much to overcome. What the search leader demonstrated was that no matter how much money is thrown to build a killer product, no amount of money in the world will ever alter human search behavior.

In 2015, Google's share of the search market, according to *NetMarketShare* (warning numbers about to appear, be prepared to reread, and I suggest that you remove the glaze from your eyes) was 66.41 percent and their share of the mobile market was 92.47 percent. Bing's share was 10.16 percent desktop, but only 1.80 percent mobile, and Yahoo!'s share was 8.76 percent desktop and 5.05 percent mobile. The mobile search figures are much more significant because more and more users are searching through mobile devices than their desktops, which are fossilizing in front of our eyes. I know what you're thinking; who are the nearly 7 percent of individuals that search on mobile devices using Yahoo! or Bing? My best guess is it's the Sherpas.

By 2016, even with Gemini fully integrated into the market, the numbers were even worse for Yahoo! Google's share of desktop search grew to 71.41 percent while their share of the mobile market increased to 94.16 percent. Yahoo!'s share decreased from 7.39 percent on desktop to 3.17 percent for mobile traffic.

There are several factors that are attributable to Google's dominance, but they all come back to a central theme. Google spent its entire existence with a singular focus on driving a search experience for users and advertisers. It developed the products it created as tools to deliver search. Its browser, Google Chrome, according to *NetMarketShare* in April 2017, had 59 percent of the installs on desktops and 54.37 percent of mobile device installs. Additionally, Google search is the default on Apple's Safari browser, which in April 2017 had a market share of 31.72 percent on mobile devices. The Google-owned Android browser market share at the same time was 6.35 percent, also on mobile devices. Therefore, 92.14 percent of the installed browsers default to the Google search engine. Hello Federal Trade Commission, U.S. Department of Justice, State Attorney Generals, European Union. Here's a hint: board game, developed by Parker Brothers in 1935, using cowboys, irons, and thimbles to travel around a nauseating square to acquire streets in Atlantic City and buy houses and hotels for 1/1,000,000,000 of 2017 housing values. (But I digress!)

As attempts to compete in social and search underwhelmed, original programming grew as a market force. Ironically, Lloyd Braun's pioneering strategy to create original programming—which failed in the mid-2000s—was finally maturing as a robust consumer choice. Unfortunately for Yahoo!, it was services like Netflix and Amazon Prime that were leading the charge. Unlike Carol Bartz or Terry Semel, Marissa Mayer was acutely aware that growing segments such as streaming video and original programming were most assuredly opportunities to look for sizable growth. Netflix debuted hit programs such as *House of Cards* and *Orange Is the New Black*. HBO unbundled its service from basic cable as viewers flocked to programs such as *Game of Thrones*. Amazon Prime introduced shows such as *Bosch* and *The Man in the High Castle*. Mayer chose to prioritize this burgeoning market.

Yahoo!'s debut program was *Community*, a comedy about the experiences of students enrolled in a Colorado community college. *Community* first aired in 2009 and ran for five seasons on NBC. It was announced that the show would air its sixth season on Yahoo!

In addition to *Community*, Yahoo! developed programs that didn't quite reach the popularity or cultural status of *Seinfeld, The Simpsons,* or *Friends.* The not-at-all-memorable programs Yahoo! produced included the very mortal *Sin City Saints* and sci-fi thriller *Other Space.* They unsurprisingly failed.

Netflix and Amazon Prime are integrated services into most smart TVs. Yahoo! isn't. While products like YouTube have significantly altered viewing habits, the best user experience for long-form programming, at least for adults, is still the big screen. Beyond the big screen, younger generations are consuming video on their mobile and tablet devices. This squeezes out the desktop, where Yahoo!'s eyeballs primarily gather. Think about it, how many people do you know who watch programming on their Dell, IBM, or HP desktop? Yahoo! was not integrated into smart TVs, and despite Marissa Mayer's Herculean efforts, the company, for the most part, disappointed in developing leading mobile- and tablet-based products.

The challenge that Marissa Mayer—or even Thomas Edison or Albert Einstein for that matter, had they come back from the dead—could not overcome was that Yahoo! would never be an effective vehicle for users to view original programming. A tired brand with a dwindling audience stuck on an early-2000s computer screen never had a chance to succeed. It's similar to sailing in the Sahara or building a solar farm in a rainforest.

Todd Spangler, a reporter for *Variety*, commented on October 27, 2015, regarding Yahoo!'s original programming efforts, "Ultimately, Yahoo didn't have the time—or the willingness to make long-term investments—to reach the scale to make ad-supported TV shows work, lacking subscription revenue or pay-TV fees the traditional ecosystem relies on. The dilemma is that had

Yahoo acquired additional shows, the endeavor might have been an even costlier debacle."

In the same article, John Blackledge, senior internet analyst at Cowen and Co. said, "TV is not their core competency, they don't have the budget to go after the best content against Netflix, Amazon, Hulu, or the networks."

The result for Yahoo! and its shareholders was a forty-two million dollar write-down on their original programming.

Not to be outdone by its failures of original programming, Yahoo! also had the chance to have another business failure in a similar but different video programming business of licensed digital streaming programming.

Yahoo! paid the NFL fifteen million dollars for the rights to be the first digital media service to live stream an NFL game. Possibly because the NFL chose to air the Bills - Jaguars as its test game, the claim of fifteen million viewers to have watched the event underwhelmed most observers. Having two teams that generate as much excitement as an Olympic curling match probably hurt the ratings as well. I'm sure the most likely explanation for its less-than-stellar debut is that football games are an event. On Sundays, homes across America fill with crazed lunatics who are screaming at their large-screen televisions as they are double fisting alcohol while they nosh on snacks. Mobile viewing has a place and is useful for viewers on the go, but football fans are so accustomed to the big screen that it will be a long wait for mobile devices to evolve as the mainstream delivery platform for NFL games.

CHAPTER 23
Activist Investor—Take 3 and Final Cut. The End of Yahoo as an Independent Company

By the end of 2015, it was apparent to everyone, including the much-maligned board, that the sand in Yahoo!'s hourglass was perilously empty. Marissa Mayer valiantly tried to rescue Yahoo!, but it was too late. Social, search, original programming, mobile, video, and every other segment were just too far out of reach for Yahoo! to catch up. They fell off the train somewhere back in 2001 (and maybe earlier), and for years tried in vain to catch up, but the train just kept pulling farther and farther away. A succession of management teams either saw the train getting closer or in some cases, hopped on a new train for a new adventure, but those hopes and dreams never materialized.

While it was clear the company was headed toward its last days as an independent organization, the Marissa Mayer years at Yahoo! were not a total washout. Yahoo!'s very shrewd investment in Alibaba and its remarkable growth from 2012 on was the cause of Yahoo! stock to rise considerably during Mayer's tenure.

From January 1, 2013, to to January 1, 2017, the stock price increased from $19.63 to $44.07. Daniel Loeb, the activist investor who hired Marissa

Mayer and brought a vision to revitalize Yahoo! rather than break up its core pieces, resigned from the board having made over one billion dollars for his Third Point Hedge Fund in just two years.

The rising stock, fueled by Alibaba, the partnership with Yahoo! Japan, and a strong cash position, gave Marissa Mayer a longer period to turn around Yahoo!'s fortune. The witching hour neared after disappointing in search, social, and original content. Yahoo! still owned about 15 percent of Alibaba, whose value in April 2016 was estimated to be between twenty-five to twenty-eight billion dollars. The value of the 35 percent ownership in Yahoo! Japan was approximately eight billion dollars, which left analysts questioning what value, if any, remained in Yahoo!'s core business.

Yahoo! reminds me of the 1960s television sitcom *Gilligan's Island*. As an eight-year-old, I watched it in syndication on a black-and-white, twenty-four inch RCA television set. I had to work the rabbit ear antennas for a clear reception. I'd be on the edge of my seat with anticipation for the first ninety percent of the program waiting for the castaways' rescue, only to be disappointed in the last two minutes when Gilligan would unwittingly blow up the liberation of Ginger and Mary Ann.

So many times, Yahoo! was close to being rescued. If only they had chosen to do paid search, if only they had acquired Google, if only they had acquired Facebook, if only they had not given search to Microsoft, if only they had embraced mobile, if only they had acquired YouTube, if only they hadn't botched Flickr, if only Tumblr hadn't failed. If only!!

Unfortunately, like Gilligan, it was always "if only." (*Gilligan's Island* was canceled before it could have a series finale, although a few years later, a made-for-television movie aired where they were rescued, and yes, I watched and cheered!)

While Yahoo!'s stock increased, shareholders had no illusions that the price was entirely indicative of the rocket fuel growth of Alibaba. Whatever new strategy Marissa Mayer and her team executed had no identifiable revenue result.

(Warning, boring numbers paragraph upcoming. Suggestion, drink a cup Dunkin' Donuts, French Vanilla Swirl before reading. Yes, a shameless plug for Dunkin' Donuts, because they are headquartered in my hometown of Canton, Massachusetts. Take that Starbucks!) In 2012, Yahoo!'s reported annual revenue of 4.98 billion dollars; in 2013, the company's annual revenue was 4.68 billion dollars; in 2014, annual revenue was 4.61 billion dollars; in 2015, annual revenue was 4.96 billion dollars; and in 2017, annual revenue was 5.16 billion dollars.

While most companies would love to have close to five billion dollars in annual revenue, Yahoo!'s numbers clearly demonstrate zero growth. Each quarter, on earnings call, the Yahoo! executive teams would report that they were executing their strategy and it would take time for the revenue to follow. Analysts are not exactly a patient bunch. Enough was enough. Unlike previous CEOs, no objective observer could ever accuse Marissa Mayer of stagnation. After years of honest attempts, it had become clear there was no magic wand that Yahoo! could wave to grow new areas of value for shareholders.

Daniel Loeb's timing was brilliant. He targeted a very unpopular board and, like others, recognized that Yahoo!'s stock was essentially a proxy for the Asian holdings. He successfully bet that it had significant room to grow. However, as the share price increased, there was little value left to ascertain. The stock, while steady, had peaked. The only possible path to unlock any additional value was to split the core operations from the Alibaba and Yahoo! Japan holdings.

For the third time since 2009, an activist investor set their arrow straight between Yahoo!'s eyes. On September 27, 2014, one week after the Alibaba IPO—which valued Yahoo!'s remaining shares at more than twenty billion dollars—Starboard Value LP announced it had acquired a significant stake in the ailing internet company. In a letter to the Yahoo! board, Jeffrey Smith, CEO of Starboard, strongly recommended a merger with AOL, suggesting cost savings of one billion dollars. Also included in the letter was an insistence that Asian assets be spun out of the core Yahoo! business. Smith forecast that unlocking the value for the company would result in a share price of fifty-six dollars. If Smith's prediction was accurate, Starboard would have generated a huge return for its investors. The stock opened at $40.41 on the next trading day after Smith's opening salvo.

In the December 3, 2014, online edition of *Fortune*, contributor William D. Cohan offered this description, "The most feared man in corporate America these days is not named Icahn, Ackman, Loeb, or Einhorn, but rather Smith, as in Jeff Smith, the boyish-looking forty-two-year-old co-founder and CEO of a previously obscure three billion dollar activist hedge fund called Starboard Value."

The comment was in reference to Smith's now legendary accomplishment of launching a proxy contest with Darden Restaurants, an 8.5-billion-dollar restaurant chain and owner of recognizable brands such as Olive Garden, Capital Grill, Seasons 52, and several others. Owning less than 10 percent of the company, Smith replaced the entire twelve-person board of directors. He presented a plan to unlock value that he believed Darden executives and the board were neglecting.

The Darden victory shot Smith to prominence. His suggestion to merge assets with AOL was not unfounded. Earlier, he had engaged in an unsuccessful activist campaign against CEO Tim Armstrong and the AOL board.

Nothing is more distracting to a company than a proxy battle. It can be counter to the primary purpose that the activist is trying to achieve, which is increased revenue leading to an increased stock price. It cripples an organization's ability to focus on improving products and services because the management team is entirely dedicated to defending the company's survival and the unpredictability deflates the entire staff. Top employees head out the door, and the ones that remain have little incentive to go "all out" each and every day.

It's similar to a basketball team where the best player is traded or leaves via free agency. The result is a string of losing seasons with no wins in sight. The confidence, willingness, and incentive for the remaining players to give maximum effort each game are destroyed slowly over time and losing becomes a habit. Team ownership tries to shake things up and recruits a coach with a proven history of winning. Unfortunately, the team chemistry is very fragile. It's fueled by wins and its tank empties with losses. After a series of losing seasons, defections of key players, and playing in front of a half-empty arena the outpouring of goodwill that existed on the day the new coach was announced dries up, and instead of just another losing season, ownership is forced to sell. When new ownership takes control, they merge the team with another losing team with no star players and re-name both teams as "Oath."

And this is by no means a complete indictment of all activist shareholders. Daniel Loeb was a positive force for Yahoo! His goal was not to destroy the company but rather to replace an ineffective management team and unproductive members of the board of directors. He successfully accomplished his goals and was rewarded handsomely for it.

Starboard had no patience to wait for a turnaround. In their defense, Marissa Mayer was given four years to execute on a growth strategy, the Asian assets of the company had grown considerably, and it was unclear where Yahoo! could establish new areas of value. They were like a boa constrictor squeezing their prey, and they weren't about to let go.

Mayer and the board also were keenly aware that the value of the company laid almost entirely with its Asian holdings. The stock which shot up between

2013 to 2015, essentially as a proxy for Alibaba, had stagnated since the IPO of the Chinese e-commerce giant. After investigating multiple paths, on January 27, 2015, the Yahoo! team announced a tax-free plan to spin off Yahoo!'s 15 percent ownership of Alibaba, by that time worth north of forty billion dollars, into a separate holding company named Aabaco.

Unfortunately for Yahoo!, Aabaco was not a slam dunk. The tax obstacles to spin off the Asian holding carried devastating consequences. For the first several months after the highly complex maneuver was announced, the markets were skeptical that it could be executed in a tax-efficient manner.

In February 2015, Yahoo! requested the opinion of the IRS as to whether the structure of Aabaco would qualify as a tax-free spin-off. As reported by CNBC on October 8, 2015, "the IRS denied the request and according to a Yahoo! Securities and Exchange Commission filing also said that the IRS had not explicitly said the proposed spin-off was taxable, leaving some hoping a deal could still be done."

The IRS denied opinion further flamed fears. On November 9, 2015, Jeffrey Smith sent a scathing letter to Marissa Mayer and Yahoo! chairman Maynard Webb. He insisted that Yahoo! scrap plans for Aabaco, and demanded that rather than spinning off the Asian assets, that they take the opposite approach of spinning off the core operation while keeping Alibaba and Yahoo! Japan stakes in an existing public company.

Smith's letter forcefully underscores in his opinion, Yahoo! is out of time and "making your [Marissa Mayer and the Yahoo! board] argument to wait for improvement [that the core business will become more valuable if management is able to turn it around] appear to be grounded more in hope than strategy."

With IRS uncertainty and shareholder upheaval, the board announced on December 8, 2015, that the Aabaco spin-off would be shelved.

On January 2, 2016, Yahoo! announced after years of resisting that its core business would be put up for sale. Unfortunately for them, nothing ever comes easy. It's ironic that even its two greatest deals, the investments in Alibaba and Yahoo! Japan, had become so complicated that it helped hasten the end of the Yahoo! as an independent, pioneering, world-changing business.

There were no shortages of suitors. On March 10, 2016, Smith—intent on having a significant voice in the sale—officially launched a proxy fight. The reality was that the sales process couldn't begin in earnest until the proxy battle was addressed and settled. Less than two months after the proxy battle was initiated the two parties settled their dispute by appointing Smith and three members of his choosing to the Yahoo! board of directors.

Three months later, on July 24, 2016, it was announced that Verizon had agreed to acquire Yahoo! for 4.8 billion dollars, officially signaling the end of Yahoo!'s twenty-two-year reign as an independent company.

Sam Ro, a reporter for Yahoo! Finance, on September 10, 2016 reported, "thirty-two parties signed confidentiality agreements with Yahoo, including ten strategic parties, and twenty-two financial sponsors."

Ro's report also revealed Yahoo! Japan entered in detailed discussions, only to be rejected by Yahoo!'s board of directors because their "proposal offered no premium for Yahoo's shares." In the end, it was apparent that Verizon's bid was the best and highest.

I was fifteen years removed from walking out of the Yahoo! offices for the last time. It was a lifetime ago. It was one child to four, car seats to driver licenses, *Goodnight Moon* to *Harry Potter*, baby cribs to California kings, merry-go-rounds to real horses, petting zoos to African safaris, training wheels to mountain bikes, miniature golf to low handicaps, daddy-daughter dances to proms, first steps to 10Ks, somersaults to back handsprings, Mommy and Me to Teen Tours in Thailand, *The Wiggles* to *Game of Thrones*, first day of kindergarten to college tours, *Finding Nemo* to *Star Wars* reboots, flip phones to iPhones, George W. Bush to Donald Trump (still can't quite believe that one), stock market bubble to housing bubble, lovable losers to Cubs World Series victors, chat rooms to Snapchat, Michael Jordan to Lebron James, Pearl Jam to Taylor Swift, gluten free to GMOs, network servers to cloud computing, and 9/11 to the Freedom Tower.

You get the point. Life changed, and it changed in a very profound and fulfilling sense. But throughout that lifetime of change, I reserved a spot deep in my heart that was always going to bleed purple and yellow. Without speaking to the Yahoo! alumni, I am confident in declaring I am not alone. Even though I knew the day was approaching when Yahoo! would be sold, and I fully anticipated Verizon would be the buyer, it still felt like a bullet had been shot at close range into my soul.

At one time or another in life, we all ask, "what if?" On average, I've asked that question every day. "What if?" questions that I've asked include: What if I had not taken this job? What if the Patriots hadn't drafted Tom Brady? What if I had just held onto that stock? What if I hadn't ever moved to New York after college? What if the Red Sox hadn't come back from a 3-0 deficit against the Yankees in the 2004 playoffs? What if I had just had the idea to market bottled water? What if someone discovers a cure for autism? I ask those and over a million other "What ifs" in my life.

An addict might ask, "What if I hadn't just tried cocaine that night at the party I wasn't even supposed to be at?" Or, "What if I had just refused a cigarette the first time it was offered?" Or, "What if I had never won playing blackjack the first time I sat at a table in Las Vegas?"

Someone sentenced to jail time might ask, "What if I had just called Uber before getting behind the wheel?" Or, "What if I had paid my taxes for the last five years?" Or, "What if I had never associated myself with that crook?"

What if an adopted child asks, "What if another couple had adopted me?"

What if a historian asks, "What if Pearl Harbor never occurred?"

What if the nation asks, "What if Donald Trump was never elected President?"

We are a world of "What ifs." It takes someone mentally invincible not to ask themselves that age-old ubiquitous question. Yahoo! has a collection of what ifs that rival any business in history. It's more than, "What if I hadn't opened that store in the wrong location?" Or, "What if I knew that employee was stealing from me earlier?" Or, "What if I had lowered my costs when my competitors did?" That's kindergarten variety to the questions of "what if" that Yahoos worldwide will have to endure forever.

What if questions all former and current Yahoo! employees will forever ask are: "What if we had just purchased Google?" "What if we had purchased Facebook?" "What if we had purchased eBay?" "What if we had purchased YouTube?" "What if we had pioneered paid search?" "What if we had accepted Microsoft's offer?" "What if we had never given up search to Bing?" "What if we had hired a technologist such as Marissa Mayer instead of a movie executive who didn't know how to use email in 2001?" "What if we could have defined ourselves earlier as either a media company or a technology company?" "What if the tech bubble had never burst?" "What if we had developed the main operating system for mobile technology?" "What if we had done more with Flickr?" "What if we had never acquired Broadcast. Com?" "What if we had a more visionary board of directors?" "What if our market value hadn't dropped by over 110 billion dollars?"

In the cemetery, Yahoo! will be buried in the same section as Blockbuster, Polaroid, and Radio Shack, with plots reserved for Sears, Levi's, and The Gap. Yahoo! will be nowhere near the section that includes Enron, Lehman

Brothers, and WorldCom, because if nothing else, Yahoo!'s integrity should never be questioned.

The words on Yahoo!'s tombstone are as follows:

"Here lies Yahoo!; once one of the most exciting pioneers that overnight changed the world for the better. A twenty-year history of poor decision-making led to the downfall of this once prodigious business leaving its thousands of dedicated Yahoo employees to ask one last time: "What if?" However much disappointment exists, we will always keep hold deep in our hearts the life-changing experience that WE WERE YAHOO!!!!"

EPILOGUE

Yahoo! was more than just a job; it was a life skills training environment for some of the brightest and most entrepreneurial individuals anywhere. The Yahoo! alumni are constantly achieving greatness. Some of the more notable ex-Yahoos include Jan Koum and Brian Acton, founders of WhatsApp, acquired by Facebook for nineteen billion dollars in February 2014. Jeff Weiner, the one-time head of Yahoo! Search, achieved great success as the CEO of LinkedIn, which in 2016 sold the company to Microsoft for twenty-six billion dollars. Andrew Braccia joined legendary Venture Capital Firm Accel Partners and became one of the first institutional investors of Facebook. Nirav Tilov co-founded NextDoor, the largest neighborhood-based social network in the world. Michael Landau co-founded Drybar, the largest "blow-dry only" salon chain in America. Stewart Butterfield founded Slack (and co-founded Flickr). Paul Graham founded Y Combinator, arguably the world's top incubator. Rob Solomon, CEO of GoFundMe and David Mandelbrot, CEO of Indigogo, are the two leading crowdfunding services in the world. Dan Rosensweig is CEO of online textbook seller Chegg. The list is extensive and includes bestselling authors Tim Sanders and Seth Godin, Dallas Mavericks owner and television personality Mark Cuban, Jeff Mallett, co-owner of the San Francisco Giants, and so many more. That is worthy of an entire book just dedicated to the success of former Yahoos.

In a July 21, 2016, article in *Fortune* about Yahoo!, alumni Dan Finnigan, a former senior vice president of Yahoo! and now president and CEO of Jobvite, was quoted as saying, "Working at Yahoo for people who were young in their career was like going to a large research university. People who started in media went into sales, people who started in sales went into recruiting, people went from engineering to product. It was highly encouraged."

I took a much different path. I left Silicon Valley and moved to Florida, where I was eventually elected to Florida State Senate. I believe I was the only ex-Yahoo to pursue a political career. (There may have been ex-Yahoos elected to local offices at a state level.) Yahoo! was my foundation for my ten-year legislative career. From the day I first campaigned to the day I reached my term limit and was out, I preached incessantly of the need for Florida to create an innovation economy along the realm of Silicon Valley. I have given speeches all over the state asking audience members to imagine what the ten-mile radius in the late nineties was like, where three companies, Google, Yahoo!, and eBay (of course, there were many others), created tens of thousands of new jobs, and its employees were so highly rewarded that the entire service economy was thrust forth. A car salesman didn't have to negotiate, an owner of a children's clothing store had lines around the corner, servers at restaurants received tips larger than they'd ever witnessed. My purpose for public service was to transport an environment of world-changing ideas to a state notable for tourism and agriculture. Without Yahoo!, I would have never created the Florida Institute for Commercialization, our statewide technology incubator, located at the University of Florida in Gainesville, that has launched over fifty technology companies since 2012. Without Yahoo!, I would never have created the Florida Opportunity Fund, which is a venture capital fund investing in companies at the first stage of institutional capital. Without Yahoo!, I would never have created the Florida Growth Fund, a billion-dollar growth stage-venture capital fund. Without Yahoo!, I never would have pushed so hard to make computer coding part of the liberal arts curriculum in high schools across the state.

Outside of the lessons of my parents, Yahoo! was the most critical education platform I've had. For me and thousands of other ex-Yahoos, our time "changing the world" has led to countless other ideas that have continued to improve the world each and every day...WE ARE YAHOO!!!

For myself, I took a much different path. I spent the ensuing years working on the four pillars I designed that day in the diner while I was being shaken down in an extortion plot.

My four pillars of life that day were spirituality, charity, children and capitalism. As humans, we evolve and almost fourteen years after that day, my life changed considerably as well. I've certainly dedicated myself to my spirituality and I've worked to improve the lives of others.

In 2006, I was elected to the Florida State Senate where I spent ten of the most fulfilling years of my life. I was proud of the legislation I authored, which improved the lives of Floridians. I passed bills that required all newborns to be checked for chronic heart ailments, I wrote laws that ensured individual with behavioral disabilities would have equal access to state jobs, I added crimes against the homeless to the state hate crime statute, I required sex offenders to include their cellphone numbers to the sex offender registry, and I played a significant role mandating that insurance companies provide coverage to individual on the autism spectrum.

The second pillar of life I pursued was creating a charity. I was the founder of SUPERB, which is an acronym for Students United with Parents and Educators to Resolve Bullying. Over fifty thousand students, mostly middle schoolers, have participated in an eight-week program I helped design to provide students with social and behavioral skills tools so they can compete in an adult world. Our program is unique in that it targets the bystander and empowers and trains that individual to intervene when they witness inappropriate behavior. A large focus is given to the child who is socially excluded. I've often said if the most popular student can sit with the child who traditionally sits alone at lunch then a life will be saved. I'm most proud of the work of SUPERB as it continues to flourish.

The third square of life I identified that day was dedication to my children. I was fortunate enough to have two amazing girls to add to our two sons at home. Like all kids, they have great days and days that they struggle, but as long as their parents are always there to provide unconditional love and unforgiving support then our kids will hopefully turn out to be their absolute best. While my children are still young, they give me the greatest joy in the world, and I hope and pray every day they are on their way to being incredibly fulfilled adults.

My final pillar of life that day was capitalism. I wanted to create the next Yahoo!, or a company that would change the world for the better. I can admit that is the one square I've yet to achieve. I've been fortunate to have some business success, primarily as the co-founder of Collegiate Images, a rights holding company for collegiate sports videos (kind of like NFL films for college), but I haven't yet created the world-changing idea that becomes part of the daily habits of every individual in the world. But I'm only just beginning part two of my life, and I have no plans of giving up checking off all four squares.

ABOUT THE AUTHOR

S enator Jeremy Ring was an early internet pioneer, as he opened the East Coast office for Yahoo! out of his apartment in early 1996. Following five extremely successful years with the company, Ring relocated his young family to Florida. In 2002, Ring founded Students United with Parents and Educators to Resolve Bullying (SUPERB). SUPERB is an eight-week program targeting middle-school students. The strategy is to identify the bystander and empower that student with bravery and confidence to intervene when inappropriate behavior is occurring. Since its inception over 50,000 students throughout the state of Florida have participated in the program.

In 2006, Ring overwhelmingly was elected to the Florida State Senate. During his ten years in office, Senator Ring introduced and passed several pieces of legislation aimed at jumpstarting the innovation economy for the state.

Additionally, Senator Ring authored several bills that gained significant national attention, including sponsoring legislation to move Florida's presidential primary to early February for the 2008 election and proposing computer programming be included as a foreign language option for high school students. Senator Ring has had thousands of mentions in all newspapers across Florida as well as appeared often as a guest on several national cable news programs on Fox, Fox Business, CNN, and NewsMax. *Time Magazine, Newsweek, USA Today, The New York Times*, and *US News* and *World Report* have all reported or opined on legislation proposed by Senator Ring. In addition, Bloomberg Business Week named Ring as one of the top ten Yahoo! alumni in America.

Senator Ring has given talks at industry and trade events, political clubs and college campuses throughout Florida. His speeches mention the rise and fall of Yahoo! and his own personal experiences helping to usher in the digital information age.